THE PAULI EXCLUSION PRINCIPLE

THE PAULI EXCLUSION PRINCIPLE

ORIGIN, VERIFICATIONS, AND APPLICATIONS

Ilya G. Kaplan

Materials Research Institute,
National Autonomous University of Mexico,
Mexico

This edition first published 2017

Registered Office
John Wiley & Sons, Ltd, The Atrium, Southern Gate, Chichester, West Sussex, PO19 8SQ, United Kingdom

For details of our global editorial offices, for customer services and for information about how to apply for permission to reuse the copyright material in this book please see our website at www.wiley.com.

Library of Congress Cataloging-in-Publication Data

Names: Kaplan, I. G. (Il′ia Grigor′evich)
Title: The Pauli exclusion principle : origin, verifications and applications / Ilya Kaplan.
Description: Chichester, West Sussex : John Wiley & Sons, Inc., 2017. |
Includes bibliographical references and index.
Identifiers: LCCN 2016018231 (print) | LCCN 2016020644 (ebook) | ISBN 9781118795323 (cloth) | ISBN 9781118795293 (epdf) | ISBN 9781118795248 (epub)
Subjects: LCSH: Pauli exclusion principle. | Quantum theory.
Classification: LCC QC174.17.P3 K37 2017 (print) | LCC QC174.17.P3 (ebook) | DDC 530.12–dc23
LC record available at https://lccn.loc.gov/2016018231

A catalogue record for this book is available from the British Library.

Cover image: David Wells/EyeEm/Gettyimages

Set in 10/12pt Times by SPi Global, Pondicherry, India

Printed in the UK

To my wife
Larisa

Contents

Preface

This book is devoted to the Pauli exclusion principle, which is a fundamental principle of quantum mechanics and has been naturally kept in mind in all its numerous applications in physics, chemistry, molecular biology, and even in astronomy, see Chapter 1. Nevertheless, to the best of my knowledge, it is the first scientific (not philosophical) book devoted to the Pauli exclusion principle. Although Wolfgang Pauli formulated his principle more than 90 years ago, its rigorous theoretical foundations are still absent. In the historical survey (Chapter 1) and in other chapters of the book, I discuss in detail still existing unsolved problems connected with the Pauli exclusion principle and for some of them suggest possible solutions.

From the beginning of my scientific activity I have been interested in the issues of symmetry in quantum mechanics and in the mathematical description of it, that is, in the group theory and, particularly, in the permutation group theory. I was impressed by the simplicity and clearness of Young's mathematical language (about Young diagrams and Young tableaux, see Appendix B), especially if one takes into account Young's individuality: most of Young's papers on the permutation group were written when he was a country parish priest. For many years I was occupied with the following question: why, according to the Pauli exclusion principle, in our Nature only the antisymmetric or symmetric permutation symmetries for identical particles are realized, while the Schrödinger equation is satisfied by functions with any permutation symmetry. The possible answers on this question I discuss in Chapter 3.

I was always impressed by the Pauli deep physical intuition, which several times struck physical community. In fact, the formulation by Pauli of his principle was to a great extent based on his intuition (it was done before the creation of modern quantum mechanics), and it gave rise to the discovery of such important quantum mechanical conception as *spin* (I describe this dramatic story in Chapter 1). Another well-known example is the Pauli prediction of neutrino. Pauli made this prediction without any experimental and theoretical indications that this chargeless and, as thought at that time, massless particle can exist. He tried to save the energy conservation law in the β-decay, because he did not agree with Niels Bohr who at that time was sure that the energy conservation law is not valid for microparticles. It turns out that Pauli was right.

The construction of functions with a given permutation symmetry is discussed in detail in Chapter 2. All necessary mathematical apparatus is given in Appendices A and B. If the total wave function of N identical particle system is represented as a product of the spatial and spin wave functions symmetrized according to the appropriate irreducible representations of the permutation group, it automatically satisfies the Pauli exclusion principle and describes the state with a definite value of the total electronic or nuclear spin.

The application of the permutation group theory for the construction of molecular wave functions makes possible elaborating effective and elegant methods for finding the Pauli-allowed states in atomic, molecular, and nuclear spectroscopy (Chapter 4). In the elaborated methods the linear groups and their interconnection with the permutation group are applied. The necessary mathematical apparatus is represented in Appendix C. The classification of the Pauli-allowed states is represented for all types of many-atom molecules with the explicit formulae for the characters of reducible representations formed by a given set of atomic states. In last sections of this chapter the methods of finding the Pauli-allowed states for an arbitrary many-particle system, containing subsystems characterized by their local symmetry, are described. These methods do not depend on the number of particles in subsystems.

Chapter 5 is devoted to exotic statistics: parastatistics and fractional statistics. Although the elementary particles obeying the parastatistics are not detected, I demonstrate that the quasiparticles (collective excitations) in a periodical lattice are obeying the modified parafermi statistics; among them are the hole pairs, which are analogue of Cooper's pairs in the high T_c superconductivity, and such well-known quasiparticles as excitons and magnons. The fractional statistics is also realized in our Nature for excitations in the fractional quantum Hall effect; these excitations can be considered as quasiparticles with fractional charge. However, the theoretical suggestions that the fractional statistics is realized in the high T_c superconductivity have not been confirmed by experiment.

I tried to write the book for a broad audience from academic researchers to graduate students connected in their work or study with quantum mechanics. Significant efforts were made to present the book so as it will be self-sufficient for readers, since all necessary apparatus of the group theory is described in the appendices.

I would like to acknowledge Lucien Piela, Lev Pitaevsky, Olga Rodimova, Oleg Vasyutinsky, Vladimir Yurovsky, and Serge Zagoulaev for useful discussions of different problems connected with the topic of the book. Special acknowledgment goes to Ulises Miranda and Alberto Lopez who helped me to correct the book.

Mexico
February 2016

1

Historical Survey

1.1 Discovery of the Pauli Exclusion Principle and Early Developments

Wolfgang Pauli formulated his principle before the creation of the contemporary quantum mechanics (1925–1927). He arrived at the formulation of this principle trying to explain regularities in the anomalous Zeeman effect in strong magnetic fields. Although in his Princeton address [1], Pauli recalled that the history of the discovery goes back to his student days in Munich. At that time the periodic system of chemical elements was well known and the series of whole numbers 2, 8, 18, 32… giving the lengths of the periods in this table was zealously discussed in Munich. A great influence on Pauli had his participation in the Niels Bohr guest lectures at Göttingen in 1922, when he met Bohr for the first time. In these lectures Bohr reported on his theoretical investigations of the Periodic System of Elements. Bohr emphasized that the question of why all electrons in an atom are not bound in the innermost shell is the fundamental problem in these studies. However, no convincing explanation for this phenomenon could be given on the basis of classical mechanics.

In his first studies Pauli was interested in the explanation of the anomalous type of splitting in the Zeeman effect in strong magnetic fields. As he recalled [1]:

> The anomalous type of splitting was especially fruitful because it exhibited beautiful and simple laws, but on the other hand it was hardly understandable, since very

> general assumptions concerning the electron using classical theory, as well as quantum theory, always led to the same triplet. A closer investigation of this problem left me with the feeling, it was even more unapproachable. A colleague who met me strolling rather aimlessly in the beautiful streets of Copenhagen said to me in a friendly manner, 'You look very unhappy'; whereupon I answered fiercely, 'How can one look happy when he is thinking about the anomalous Zeeman effect?'

Pauli decided to analyze the simplest case, the doublet structure of the alkali spectra. In December 1924 Pauli submitted a paper on the Zeeman effect [2], in which he showed that Bohr's theory of doublet structure based on the nonvanishing angular moment of a closed shell, such as K-shell of the alkali atoms, is incorrect and closed shell has no angular and magnetic moments. Pauli came to the conclusion that instead of the angular momentum of the closed shells of the atomic core, a new quantum property of the electron had to be introduced. In that paper he wrote, remarkable for that time, prophetic words. Namely:

> According to this point of view, the doublet structure of alkali spectra ... is due to a particular two-valuedness of the quantum theoretic properties of the electron, which cannot be described from the classical point of view.

This nonclassical *two-valued nature of electron* is now called *spin*. In anticipating the quantum nature of the magnetic moment of electron before the creation of quantum mechanics, Pauli exhibited a striking intuition.

After that, practically all was ready for the formulation of the exclusion principle. Pauli also stressed the importance of the paper by Stoner [3], which appeared right at the time of his thinking on the problem. Stoner noted that the number of energy levels of a single electron in the alkali metal spectra for the given value of the principal quantum number in an external magnetic field is the same as the number of electrons in the closed shell of the rare gas atoms corresponding to this quantum number. On the basis of his previous results on the classification of spectral terms in a strong magnetic field, Pauli came to the conclusion that a single electron must occupy an entirely nondegenerate energy level [1].

In the paper submitted for publication on January 16, 1925 Pauli formulated his principle as follows [4]:

> In an atom there cannot be two or more equivalent electrons, for which in strong fields the values of all four quantum numbers coincide. If an electron exists in an atom for which all of these numbers have definite values, then this state is 'occupied.'

In this paper Pauli explained the meaning of four quantum numbers of a single electron in an atom, n, l, $j = l \pm 1/2$, and m_j (in the modern notations); by n and l he denoted the well known at that time the principal and angular momentum quantum numbers, by j and m_j—the total angular momentum and its projection, respectively. Thus, Pauli characterized the electron by some additional quantum number j, which in the case of $l=0$ was equal to $\pm 1/2$. For the fourth quantum number of the electron Pauli did not give any physical interpretations, since he was sure, as we discussed above, that it cannot be described in terms of classical physics.

Introducing two additional possibilities for electron states, Pauli obtained $2(2l+1)$ possibilities for the set (n, l, j, m_j). That led to the correct numbers 2, 8, 18, and 32 for the lengths of the periods in the Periodic Table of the Elements.

As Pauli noted in his Nobel Prize lecture [5]: "...physicists found it difficult to understand the exclusion principle, since no meaning in terms of a model was given to the fourth degree of freedom of the electron." Although not all physicists! Young scientists first Ralph Kronig and then George Uhlenbeck and Samuel Goudsmit did not take into account the Pauli words that the electron fourth degree of freedom cannot be described by classical physics and suggested the classical model of the spinning electron. Below I will describe in some detail the discovery of spin using the reminiscences of the main participants of this dramatic story.

Kronig recalled [6] that on January 7, 1925, at the age of 20, he, as a traveling fellow of the Columbia University, arrived in the small German university town of Tübingen to see Landé and Gerlach. At the Institute of Physics Kronig was received by Landé with the remark that it was a very opportune moment, since he was expecting Pauli the following day and he just received a long and very interesting letter from Pauli. In that letter Pauli described his exclusion principle. Pauli's letter made a great impression on Kronig and it immediately occurred to him that additional to the orbital angular momentum l the momentum $s=1/2$ can be considered as an intrinsic angular momentum of the electron. The same day Kronig performed calculations of the doublet splitting. The results encouraged him, although the obtained splitting was too large, by a factor of 2. He reported his results to Landé. Landé recommended telling these results to Pauli. Next day Pauli arrived at Tübingen, and Kronig had an opportunity to discuss with him his ideas. As Kronig [6] wrote: "Pauli remarked: '*Das ist ja ein ganz Einfall*',[1] but did not believe that the suggestion had any connection with reality."

Later Kronig discussed his ideas in Copenhagen with Heisenberg, Kramers, and others and they also did not approve them. Under the impression of the negative reaction of most authoritative physicists and some serious problems in his calculations Kronig did not publish his ideas about a spinning electron. In the letter to van der Waerden [7] Kronig wrote about the difficulties he met in his studies of the spinning electron:

> First, the factor 2 already mentioned. Next, the difficulty to understand how a rotation of the electron about its axis would yield a magnetic moment of just one magneton. Next, the necessity to assume, for the rotating charge of an electron of classical size, velocities

[1] This is a very funny idea.

> surpassing the velocity of light. Finally, the smallness of the magnetic moments of atomic nuclei, which were supposed, at that time, to consist of proton and electrons

Independent of Kronig, the Dutch physicists Uhlenbeck and Goudsmit after reading the Pauli paper on his exclusive principle also arrived at the idea of the spinning electron. In his address, delivered at Leiden on the occasion of his Lorentz Professorship, Uhlenbeck [8] told in detail the story of their discovery and its publication.[2]

According to Uhlenbeck, he and Goudsmit were greatly affected by the Pauli exclusion principle, in particular by the fourth quantum number of the electron. It was a mystery, why Pauli did not suggest any concrete picture for it. Due to their conviction that every quantum number corresponds to a degree of freedom, they decided that the point model for the electron, which had only three degrees of freedom, was not appropriate and the electron should be assumed as a small sphere that could rotate. However, very soon they recognized that the rotational velocity at the surface of the electron had to be many times larger than the velocity of light. As Uhlenbeck writes further,

> ...we had not the slightest intention of publishing anything. It seems so speculative and bold, that something ought to be wrong with it, especially since Bohr, Heisenberg and Pauli, our great authorities, had never proposed anything of this kind. But of course we told Erenfest. He was impressed at once, mainly, I feel, because of the visual character of our hypothesis, which was very much in his line. ... and finally said that it was either highly important or nonsense, and that we should write a short note for Naturwissenschaften and give it to him. He ended with the words 'und dann werden wir Herrn–Lorentz fragen'.[3] This was done. ... already next week he (Lorentz) gave us a manuscript, written in his beautiful hand writing, containing long calculations on the electromagnetic properties of rotating electrons. We could not fully understand it, but it was quite clear that the picture of the rotating electron, if taken seriously, would give rise to serious difficulties. ... Goudsmit and myself felt that it might be better for present not to publish anything; but when we said this to Erenfest, he answered: 'Ich habe Ihren Brief schon längst abgesandt; Sie sind beide jung genug um sich eine Dummheit leisten zu können.[4]

Thus, the short letter of Uhlenbeck and Goudsmit was transmitted by Erenfest to the editor of *Naturwissenschaften* and soon published [9]. Then in February 1926 they published a paper in *Nature* [10]. In the letter to Goudsmit from November 21, 1925 (see van der Waerden [7]), Heisenberg congratulated him with their paper but also asked him how he envisaged getting rid of the wrong factor 2 in the doublet splitting formula. Bohr, who was initially rather skeptic about the hypothesis of the spinning electron and did not approve the Kronig idea, gradually changed his mind.

---

[2] English translation of an essential part of Uhlenbeck's address represented in Ref. [7].

[3] ...and then we will also ask Mr. Lorentz.

[4] I have already sent your letter some time ago. You are both young enough and can afford yourself a foolishness.

The meeting with Einstein became crucial. In his letter to Kronig from March 26, 1926 (see van der Waerden [7]), Bohr writes:

> When I came to Leiden to the Lorenz festivals (December 1925), Einstein asked the very first moment I saw him what I believe about the spinning electron. Upon my question about the cause of the necessary mutual coupling between spin axis and the orbital motion, he explained that this coupling was an immediate consequence of the theory of relativity. This remark acted as a complete revelation to me, and I have never since faltered in my conviction that we at last were at the end of our sorrows.

Under the influence of Bohr's opinion on the idea of spinning electron, Heisenberg at last removed his objections.

However, Pauli did not! His deep intuition did not allow him at once to admit the hypothesis of the spin as an intrinsic angular momentum of the rotating electron. Pauli's objections resulted from the wrong factor 2 in the doublet splitting, but mainly from the classical nature of the spin hypothesis. After the Lorentz festival (December 1925), Pauli met Bohr in Berlin and in strong words expressed his dissatisfaction that Bohr changed his position. Pauli was convinced that a new "Irrlehre"[5] has arisen in atomic physics, as van der Waerden wrote in his recollections [7].

Meanwhile, in April 1926, a young English physicist Llewellyn Thomas, who had spent half a year in Copenhagen with Bohr, published a letter in *Nature* [11], where he presented a relativistic calculation of the doublet splitting. Thomas demonstrated that the wrong factor 2 disappears and the relativistic doublet splitting does not involve any discrepancy. In the end Thomas noted, "... as Dr. Pauli and Dr. Heisenberg have kindly communicated in letters to Prof. Bohr, it seems possible to treat the doublet separation as well as the anomalous Zeeman effect rigorously on the basis of the new quantum mechanics." Thus, this time Pauli was certain that the problem can be treated rigorously by the quantum mechanical approach. The relativistic calculations by Thomas finally deleted all his doubts.

In his Nobel Prize lecture Pauli recalled [5]:

> Although at first I strongly doubted the correctness of this idea because of its classical mechanical character, I was finally converted to it by Thomas [11] calculations on the magnitude of doublet splitting. On the other hand, my earlier doubts as well as the cautious expression 'classically non-describable two-valuedness' experienced a certain verification during later developments, as Bohr was able to show on the basis of wave mechanics that the electron spin cannot be measured by classically describable experiments (as, for instance, deflection of molecular beams in external electromagnetic fields) and must therefore be considered as an essentially quantum mechanical property of the electron.

[5] Heresy.

It is now clear that Pauli was right in not agreeing with the classical interpretation of the fourth degree of freedom. The spin in principle cannot be described by classical physics. The first studies devoted to applying the newborn quantum mechanics to many-particle systems were performed independently by Heisenberg [12] and Dirac [13]. In these studies, the Pauli principle, formulated as the prohibition for two electrons to occupy the same quantum state, was obtained as a consequence of the antisymmetry of the wave function of the system of electrons.

It is instructive to stress how young were the main participants of this dramatic story. They were between 20 and 25 years. In 1925, the creators of quantum mechanics—Werner Heisenberg (1901–1976), Paul Dirac (1902–1984), Wolfgang Pauli (1900–1960), Enrico Fermi (1901–1954), and some others—were of the same age. Namely: Heisenberg—24, Dirac—23, Pauli—25, Fermi—24.

* *
*

In his first paper [12], submitted in June 1926, Heisenberg constructed the antisymmetric Schrödinger eigenfunction for the system of n identical particles (electrons) as a sum:

$$\varphi = \frac{1}{\sqrt{n!}} \sum (-1)^{\delta_k} \varphi_1 \left(m_\alpha^k \right) \varphi_2 \left(m_\beta^k \right) \ldots \varphi_n \left(m_\nu^k \right) \tag{1.1}$$

where δ_k is a number of transpositions in a permutation, P_k (a parity of permutation), and $m_\alpha^k m_\beta^k \ldots m_\nu^k$ the new order of quantum numbers $m_1 m_2 \ldots m_n$ after the application of permutation P_k. Heisenberg concluded that this function cannot have two particles in the same state, that is, it satisfies the Pauli exclusion principle. In the following paper [14], submitted in July 1926, Heisenberg considered a two-electron atom and from the beginning assumed that the Pauli-allowed wave functions must be antisymmetric. He demonstrated that the total antisymmetric wave function can be constructed as a product of spatial and spin wave functions and discussed two possibilities: A—the symmetric eigenfunction of the space coordinates is multiplied by the antisymmetric eigenfunction of the spin coordinates; B—the antisymmetric eigenfunction of the space coordinates is multiplied by the symmetric eigenfunction of the spin coordinates. Case A corresponds to the atomic singlet state with the total spin $S = 0$; case B corresponds to the triplet state with $S = 1$. Heisenberg presented detailed calculations for the atom He and the ion Li^+. These were first quantum mechanical calculations of the atomic states characterized by the total spin S of the atom defined by the vector addition of the spins of the individual electrons.

Dirac [13] began with the two-electron atom and noted that the states differing by permutations of electrons $\psi_n(1)\psi_m(2)$ and $\psi_n(2)\psi_m(1)$ correspond to the same state of the atom; these two independent eigenfunctions must give rise to the symmetric and antisymmetric linear combinations providing a complete solution of the two-electron problem. Then Dirac considered the systems with any number of electrons and represents an N-electron antisymmetric function as a determinant[6]:

$$\begin{vmatrix} \psi_{n_1}(1) & \psi_{n_1}(2) & \ldots & \psi_{n_1}(r) \\ \psi_{n_2}(1) & \psi_{n_2}(2) & \ldots & \psi_{n_2}(r) \\ \ldots & \ldots & \ldots & \ldots \\ \psi_{n_r}(1) & \psi_{n_r}(2) & \ldots & \psi_{n_r}(r) \end{vmatrix} \quad (1.2)$$

After presenting the many-electron wave function in the determinantal form Dirac wrote: "An antisymmetrical eigenfunction vanishes identically when two of the electrons are in the same orbit. This means that in the solution of the problem with antisymmetrical eigenfunctions there can be no stationary states with two or more electrons in the same orbit, which is just Pauli's exclusion principle. The solution with symmetrical eigenfunctions, on the other hand, allows any number of electrons to be in the same orbit, so that this solution cannot be the correct one for the problem of electrons in an atom."

In the second part of his paper [13], Dirac considered an assembly of noninteracting molecules. At that time it was supposed that molecules are resembled electrons and should satisfy the Pauli exclusion principle. Dirac described this assembly, in which every quantum state can be occupied by only one molecule, by the antisymmetric wave functions and obtained the distribution function and some statistical quantities. It should be mentioned that these statistical formulae were independently published by Fermi [16] in the paper submitted several months earlier than the Dirac paper [13]. Fermi also considered an assembly of molecules and although his study was performed within the framework of classical mechanics, the results were the same as those obtained by Dirac who applied the newborn quantum mechanics. This concluded the creation of the statistics, which is at present named the *Fermi–Dirac statistics*.

In the same fundamental paper [13], Dirac considered the assembly described by the symmetric wave functions and concluded that he arrived at the already known Bose–Einstein *statistical mechanics*.[7] Dirac stressed that the light quanta must be described by the symmetric wave functions and he specially noted that a system of electrons cannot be described by the symmetric wave functions since this allows any number of electrons to occupy a quantum state.

---

[6] It is important to note that the determinantal representation of the electronic wave function, at present widely used in atomic and molecular calculations, was first introduced in general form by Dirac [13] in 1926. In 1929, Slater [15] introduced the spin functions into the determinant and used the determinantal representation of the electronic wave function (so-called Slater's determinants) for calculations of the atomic multiplets.

[7] This statistics was introduced for the quanta of light by Bose [17] and generalized for particles by Einstein [18, 19].

Thus, with the creation of quantum mechanics, the prohibition on the occupation numbers of electron system states was supplemented by the prohibition of all types of permutation symmetry of electron wave functions except for antisymmetric ones.

The first quantum mechanical calculation of the doublet splitting and the anomalous Zeeman effect for atoms with one valence electron was performed by Heisenberg and Jordan [20] in 1926. They used the Heisenberg matrix approach and introduced the spin vector $\boldsymbol{s}$ with components s_x, s_y, and s_z with commutations relations the same as for the components of the orbital angular moment $\boldsymbol{l}$. The spin–orbit interaction was taken as proportional to $\boldsymbol{l}{\cdot}\boldsymbol{s}$. The application of the perturbation theory led to results, which were in full accordance with experiment.

In 1927, Pauli [21] studied the spin problem using the wave functions. Pauli introduced the spin operators $\boldsymbol{s}_x, \boldsymbol{s}_y, \boldsymbol{s}_z$ acting on the wave functions, which depend on the three spatial coordinates, q, and a spin coordinate. Pauli took s_z as a spin coordinate. The latter is discrete with only two values. Therefore, the wave function $\psi(q, s_z)$ can be presented as a two-component function with components $\psi_\alpha(q)$ and $\psi_\beta(q)$ corresponding to $s_z = 1/2$ and $s_z = -1/2$, respectively. The operator, acting on the two-component functions, can be presented as a matrix of the second order. Pauli obtained an explicit form of the spin operators, representing them as $\boldsymbol{s}_x = 1/2\boldsymbol{\sigma}_x$, $\boldsymbol{s}_y = 1/2\boldsymbol{\sigma}_y$, and $\boldsymbol{s}_z = 1/2\boldsymbol{\sigma}_z$, where $\boldsymbol{\sigma}_\tau$ are the famous *Pauli matrices*:

$$\sigma_x = \begin{pmatrix} 0 & 1 \\ 1 & 0 \end{pmatrix}, \quad \sigma_y = \begin{pmatrix} 0 & -i \\ i & 0 \end{pmatrix}, \quad \sigma_z = \begin{pmatrix} 1 & 0 \\ 0 & -1 \end{pmatrix}. \tag{1.3}$$

Applying his formalism to the problem of the doublet splitting and the anomalous Zeeman effect, Pauli obtained, as can be expected, the same results as Heisenberg and Jordan [20] obtained by the matrix approach.

The Pauli matrices were used by Dirac in his derivation of the Schrödinger equation for the relativistic electron [22]. However, for most of physicists the two-component functions that do not transform like vectors or tensors seemed very strange. As van der Waerden recalled [7]: "Erenfest called these quantities *Spinors* and asked me on his visit to Göttingen (summer, 1929): 'Does a Spinor Analysis exist, which every physicist can learn like Tensor Analysis, and by which all possible kinds of spinors and all invariant equations between spinors can be written down?' " This request made by an outstanding physicist was fulfilled by van der Waerden in his publication [23].

After these publications, the first stage of the quantum mechanical foundation of the Pauli exclusion principle and the conception of the spin could be considered as completed. Although it is necessary to mention very important applications of the group-theoretical methods to the quantum mechanical problems, which were developed at that time by John von Neumann and Eugene Wigner [24–27]. Very soon the three remarkable books on the group theory and quantum mechanics were

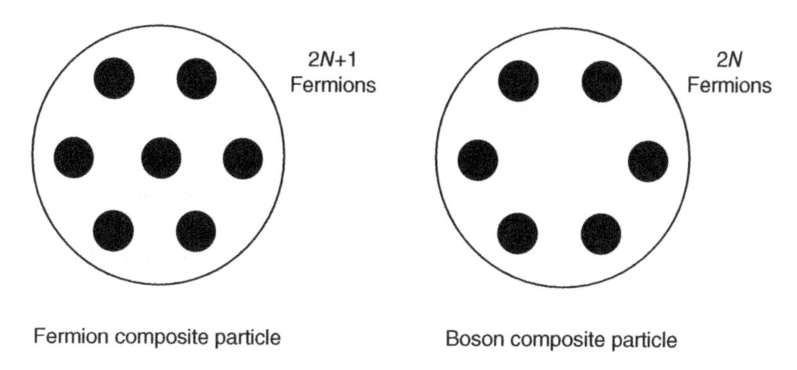

Fig. 1.1 The statistics of composite particles

published; first by Herman Weyl [28] and then by Wigner [29] and by van der Waerden [30].

The discovery of various types of elementary particles in the 1930s allowed formulating the Pauli exclusion principle in a quite general form. Namely:

The only possible states of a system of identical particles possessing spin s *are those for which the total wave function transforms upon interchange of any two particles as*

$$P_{ij}\Psi(1,\ldots,i,\ldots j,\ldots,N) = (-1)^{2s}\Psi(1,\ldots,i,\ldots j,\ldots,N), \tag{1.4}$$

that is, it is symmetric for the integer values of s *(the Bose–Einstein statistics) and antisymmetric for the half-integers (the Fermi–Dirac statistics).*

The Pauli exclusion principle formulated above also holds for composite particles. First, it was discussed by Wigner [31] and independently by Ehrenfest and Oppenheimer [32]. The latter authors considered some clusters of electrons and protons; it can be atoms, molecules, or nuclei (at that time the neutron had not been discovered yet and it was believed that the nuclei were built from electrons and protons). They formulated a rule, according to which statistics of a cluster depends upon the number of particles from which they are built up. In the case of odd number of particles it is the Fermi–Dirac statistics, while in the case of even number it is the Bose–Einstein statistics, see Fig. 1.1. It was stressed that this rule is valid, if the interaction between composite particles does not change their internal states; that is, the composite particle is stable enough to preserve its identity.

A good example of such stable composite particle is the atomic nucleus. It consists of nucleons: protons and neutrons, which are fermions because they both have $s = 1/2$. Depending on the value of the total nuclear spin, one can speak of boson nuclei or fermion nuclei. The nuclei with an even number of nucleons have an integer value of the total spin S and are bosons; the nuclei with an odd number of nucleons have a half-integer value of the total spin S and are fermions.

A well-known example, in which the validity of the Pauli exclusion principle for composite particles can be precisely checked in experiment, is the $^{16}O_2$ molecule. The nucleus ^{16}O is a boson composite particle, so the total wave function of the $^{16}O_2$ molecule must be symmetric under the permutations of nuclei. At the Born–Oppenheimer approximation [33] a molecular wave function can be represented as a product of the electronic, Ψ_{el}, and nuclear, Φ_n, wave functions. At the equilibrium distances the nuclear wave function, in its turn, can be represented as a product of the vibrational, Φ_{vib}, and rotational, Φ_{rot}, wave functions. Thus,

$$\Psi\left({}^{16}\mathrm{O}_a - {}^{16}\mathrm{O}_b\right) = \Psi_{\mathrm{el}}(ab)\Phi_{\mathrm{vib}}(ab)\Phi_{\mathrm{rot}}(ab). \tag{1.5}$$

The vibrational wave function, $\Phi_{\mathrm{vib}}(ab)$, depends only on the magnitude of the interatomic distance and remains unaltered under the interchange of the nuclei. The ground state electronic wave, $\Psi_{\mathrm{el}}(ab)$, is antisymmetric under the interchange of the nuclei. Hence, for fulfilling the boson symmetry of the total wave function (1.5), the rotational wave function, $\Phi_{\mathrm{rot}}(ab)$, must be also antisymmetric under the interchange of the nuclei. The symmetry of the rotational wave function in the state with the rotational angular momentum K is determined by the factor $(-1)^K$. Therefore, in the ground electronic state the even values of K are forbidden and only odd values of K are allowed. Exactly this was revealed in 1927 in spectroscopic measurements [34] made before the theoretical studies [31, 32].

I presented above the general formulation of the Pauli exclusion principle in the terms of the permutation symmetry of the total wave function. There is also a formulation of the Pauli exclusion principle in the second quantization formalism. The second quantization for the electromagnetic field, that is, for bosons, was created by Dirac [35]; the commutations relations for fermion and boson operators in the explicit modern form were formulated by Jordan and Wigner [36], see also references therein.

For bosons, which are described by the symmetric wave functions and satisfy the Bose–Einstein statistics, the commutation relations for the creation b_k^+ and annihilation b_k operators in the quantum state k are (see Appendix E)

$$\begin{aligned} \left[b_k, b_{k'}^+\right]_- &= b_k b_{k'}^+ - b_{k'}^+ b_k = \delta_{kk'}, \\ \left[b_k, b_{k'}\right]_- &= \left[b_k^+, b_{k'}^+\right]_- = 0, \end{aligned} \tag{1.6}$$

while for fermions, which correspond to the Fermi–Dirac statistics with the antisymmetric wave functions, the commutation relations for the creation c_k^+ and annihilation c_k operators (in the fermion case they are transformed to the anticommutation relations) are

$$\begin{aligned} \left[c_k, c_{k'}^+\right]_+ &= c_k c_{k'}^+ + c_{k'}^+ c_k = \delta_{kk'}, \\ \left[c_k, c_{k'}\right]_+ &= \left[c_k^+, c_{k'}^+\right]_+ = 0. \end{aligned} \tag{1.7}$$

As follows from the second line of the fermion anticommutation relations (1.7),

$$\left(c_k^+\right)^2 = 0, \tag{1.8}$$

or no more than one fermion particle can be created in one quantum state, which is exactly the primary formulation of the Pauli principle. A more detailed description of the second quantization formalism is presented in Appendix E.

Some of the field theory specialists claimed that the second quantization formulation of the Pauli exclusion principle is the most general; see, for instance, Ref. [37]. I do not think so, these formulations are quite different. On the one hand, the second quantization formalism is developed for N-particle system in the case when each particle is characterized by its own wave function (so-called one-particle approximation),[8] while the ψ-formalism considers the permutation symmetry of the total wave function in any approximation, even for an exact solution when the particles lost their individualities. Thus, in this sense the ψ-formulation of the Pauli exclusion principle is more general than the formulation in the second quantization formalism. On the other hand, for the composite particles the formulation in the second quantization formalism allows to take into account the internal structure of the composite particle. The symmetry of the wave functions of N-particle system does not change when we go from elementary to composite particles satisfying the same statistics, while for the commutation relations of the second quantization operators it is not true; in the case of composite particles they are changed. We will discuss this problem and the reasons for this in the next subsection.

1.2 Further Developments and Still Existing Problems

In 1932, Chadwick [38] discovered neutron. In the same year, Heisenberg [39] considered consequences of the model, in which the nuclei are built from protons and neutrons, assuming that the forces between all pairs of particles are equal and in this sense the proton and neutron can be considered as different states of one particle. Heisenberg [39] introduced a variable τ. The value $\tau = -1$ was assigned to the proton state, and the value $\tau = 1$ to the neutron state. Wigner [40] called τ as *isotopic spin* (at present named also as *isobaric spin*). Taking into account for protons and neutrons their nuclear spin $s = 1/2$ too, Wigner studied the nuclear charge-spin supermultiplets for Hamiltonian not involving the isotope spin and the ordinary spin as well, see also Refs. [41, 42].

In the 1940s, Giulio Racah published a series of four papers [43–46], in which he considerably improved methods of classification and calculation of atomic spectra. At that time the calculations of atomic spectra were performed by the diagonal-sum

[8] It is natural in the relativistic theory where the number of particles in the system can be changed.

procedure elaborated in 1929 by Slater [15]; its generalized version extended to electron shells up to f electrons was represented in the widely used Condon and Shortley book [47]. The calculations by the diagonal-sum method were very lengthy and did not give general formulas, but only numerical tables.

Racah [43–46] developed new elegant and effective methods introducing in the atomic spectroscopy the tensor operator techniques and the concept of the *fractional parentage coefficients*. The latter permitted a genealogical construction of the N-electron wave function from the parent (N–1)-electron states. The antisymmetric wave function for the configuration ℓ^N was presented as a linear combination of the wave functions obtained by the addition of an electron with the angular momentum ℓ to the possible states of the configuration ℓ^{N-1}. Racah studied the transformation matrices for the three-dimensional rotation group $\mathbf{R}_3$ connecting different coupling schemes for three angular momenta and introduced so-called Racah's W coefficients, see Appendix C, Section C.2. In the last paper [46] he applied the theory of continuous group to the problem of classification of the Pauli-allowed states for configurations of equivalent electrons. These publications made a great impact on the atomic spectroscopy as well as on the nuclear physics.

In 1950, Jahn [48] used the Racah approach for a classification, in the Russell–Saunders (LS) coupling scheme, of the states for the nuclear d-shell according to their transformation properties under the group of rotations in the five-dimensional space of the orbital states of the d-particle. He determined the charge–spin structure of all Wigner's supermultiplets [40]. Then Jahn with coauthors calculated the energy of nuclear d- and p-shells at the Hartree–Fock approximation using the method of the fractional parentage coefficients [49–52]. The new point in these studies was the presentation of the total wave function as a linear combination of the products of orbital and charge–spin wave functions symmetrized according to the mutually conjugate representations $\Gamma^{[\lambda]}$ and $\Gamma^{[\tilde{\lambda}]}$ of the permutation group that provides the antisymmetry of the total wave function. The Young diagram $[\tilde{\lambda}]$ is dual to $[\lambda]$, that is, it is obtained from the latter by replacing rows by columns, see Appendix A. For jj-coupling in nuclear shell model this approach was elaborated in Refs. [53, 54]. In many problems, in particular for the classification of the Pauli-allowed states, an employment of the permutation group proved to be more effective than the original Racah approach. This was demonstrated in the nuclear studies cited above and in our studies [55–57] devoted to the application of the permutation group apparatus to molecular spectroscopy for finding nuclear and electronic multiplets allowed by the Pauli exclusion principle.

In 1961, Kaplan [58] introduced the transformation matrices for the permutation group connecting representations with different types of reduction on subgroups, which can be considered as an analog of the transformation matrices connecting different angular momentum couplings for the rotation group. The symmetry properties of the transformation matrices for the permutation group were studied by Kramer [59, 60] who showed that these matrices are identical with the invariants of the unitary groups. In the case of electrons the transformation matrices for the permutation group can be expressed in terms of the invariants of the group SU_2, which are just the $3nj$-symbols of the three-dimensional rotation group; these connections are also discussed in a special section "The Kaplan matrices and nj-symbols for group SU(n)" in review by Neudachin et al. [61]. Employment of Kaplan's matrices allowed obtaining the general expressions for the spatial-coordinate fractional parentage coefficients of an arbitrary multishell nuclear (or atomic) configuration [62].

The first application of the permutation group to molecular problems was done by Kotani and Siga [63] to study the CH_4 molecule. Then Kotani and coworkers [64, 65] applied this approach to the configuration interaction calculations of diatomic molecules. In 1963, Kaplan [66, 67] applied his methodology [58, 62] developed for the spherical symmetry case to molecular systems and then elaborated it in a series of papers [68], where this approach was named as the coordinate (that meant spatial coordinate) function method. Later on it was named as *spin-free quantum chemistry*. These studies were systemized and generalized in a monograph [69].

Though the concept of the spin has enabled to explain the nature of the chemical bond, electron spins are not involved directly in the formation of the latter. The interactions responsible for chemical bonding have a purely electrostatic nature. The main idea of the spin-free quantum chemistry is to use in the calculations with a nonrelativistic Hamiltonian only the spatial wave functions $\Phi_r^{[\lambda]}$, which entered the total wave function as products with the spin wave functions $\Omega_{\tilde{r}}^{[\tilde{\lambda}]}$. The Young diagram $[\tilde{\lambda}]$ is dual to $[\lambda]$, which provides the antisymmetry of the total wave function. For the spin $s = 1/2$, the spin Young diagram $[\tilde{\lambda}]$ is uniquely connected with the value of the total spin S. Thus, the spatial wave functions $\Phi_r^{[\lambda]}$ describe the antisymmetric state with the definite total spin S. In the case of the Hamiltonian not containing spin-dependent interactions, this approach is more natural than the employment of the Slater determinants; it allows obtaining the energy matrix elements in an explicit compact form for arbitrary electronic configurations in the state with a definite spin S [69, 70].

At the same time the concepts of the spin-free quantum chemistry were independently developed by Matsen [71–74]. Later this approach was applied to molecular calculations by Goddard [75, 76], Gallup [77, 78], Gerratt [79, 80], and many others.

As we mentioned in the end of the previous subsection, despite the fact that the wave function of composite particles can be characterized only by the boson or fermion permutation symmetry, its second quantization operators do not obey the pure boson or fermion commutation relations. This was studied in detail by Girardeau [81, 82] and Gilbert [83], see also recent publications [84, 85]. When the internal structure of the composite particle is taken into account, the deviations from the purely bosonic or fermionic properties usually appear. For two fermions it was revealed earlier, in 1957, in the Bardeen–Cooper–Schrieffe (BCS) theory of superconductivity [86] based on the conception of the Cooper pairs [87], see an excellent description of their theory by Schrieffer [88].

The operators of creation, $b_{\mathbf{k}}^{+}$, and annihilation, $b_{\mathbf{k}}$, of Cooper's pair in a state $(\mathbf{k}\alpha, -\mathbf{k}\beta)$, where $\mathbf{k}$ is the electron momentum and α and β are the spin projections, are defined as products of the electron creation, $c_{\mathbf{k}\alpha}^{+}$, and annihilation, $c_{\mathbf{k}\alpha}$, operators

$$b_{\mathbf{k}}^{+} = c_{\mathbf{k}\alpha}^{+} c_{-\mathbf{k}\beta}^{+}, \quad b_{\mathbf{k}} = c_{-\mathbf{k}\beta} c_{\mathbf{k}\alpha}. \tag{1.9}$$

The Cooper pairs have spin $S=0$, so the permutation symmetry of their wave functions is bosonic. But their operators do not obey the boson commutation relations. Direct calculation results in

$$\left[b_{\mathbf{k}}, b_{\mathbf{k}'}^{+}\right]_{-} = \delta_{\mathbf{k}\mathbf{k}'}\left(1 - c_{\mathbf{k}\alpha}^{+} c_{\mathbf{k}\alpha} - c_{\mathbf{k}\beta}^{+} c_{\mathbf{k}\beta}\right). \tag{1.10}$$

Only in the case $\mathbf{k} \neq \mathbf{k}'$

$$\left[b_{\mathbf{k}}, b_{\mathbf{k}'}\right]_{-} = 0 \tag{1.11}$$

and Cooper's pairs obey the boson commutation relations. For $\mathbf{k} = \mathbf{k}'$

$$\left[b_{\mathbf{k}}, b_{\mathbf{k}}\right]_{-} = 1 - c_{\mathbf{k}\alpha}^{+} c_{\mathbf{k}\alpha} - c_{\mathbf{k}\beta}^{+} c_{\mathbf{k}\beta}. \tag{1.12}$$

Due to the fermion nature of electrons, the commutation relations for Cooper's pair operators are not bosonic and even more, they have the fermion occupation numbers for one-particle states, since

$$\left(b_{\mathbf{k}}^{+}\right)^{2} = \left(b_{\mathbf{k}}\right)^{2} = 0. \tag{1.13}$$

Thus, Cooper's pair operators may not be considered as the Bose, or as the Fermi operators [89, 90], see Section 5.3.

One of the first studies of the effective repulsion between identical elementary particles was performed by Zeldovich [91] who considered the repulsion between elementary barions (neutrons, protons, and others) and showed that the Pauli repulsion arises when the overlap of wave functions become appreciable. This corresponds to the well known in atomic and molecular physics *exchange interaction* stipulated by the requirement of the antisymmetry of many-electron wave functions. The exchange interaction is a direct consequence of the Pauli exclusion principle. For taking into account the Pauli repulsion, different computational schemes were elaborated. We mention two methods: the Pauli repulsion operator [92], see recent application in Ref. [93], and the so-called Pauli blockade method [94], see recent application in Ref. [95].

However, the first application of the Pauli repulsion was performed by Fowler [96] in astrophysics already in the next year after Pauli suggested his principle. Fowler applied the Pauli exclusion principle for an explanation of the white-dwarf structure. The radius of the white dwarfs is comparable with the earth's radius, while their mass is comparable with the solar mass. Therefore, the average density of the white dwarfs is 10^6 times greater than the average density of the sun; it is approximately 10^6 g/cm^3. The white dwarfs are composed from plasma of bare nuclei and electrons. Fowler [96] had resolved a paradox: why such dense objects, as the white dwarfs, are not collapsed at low temperature? He applied to the electron gas in the white dwarfs the Fermi–Dirac statistics, introduced in the same 1926, and showed that even at very low temperatures the electron gas, called at this conditions as degenerate, still possesses a high energy; compressing of a white dwarf leads to increase of the inner electron pressure. Fowler concluded his paper in the following manner: "...the origin of this important part of interatomic forces (*repulsion, IK*) is ... in the quasi-thermodynamic consequences of the existence of the quantum constraints embodied in Pauli's principle." Thus, the Pauli repulsion prevents the white dwarfs from the gravitational collapse.

In Refs. [85, 97] the Pauli exclusion principle was connected with such interesting and mysterious quantum phenomenon as entanglement [98], which at present is broadly implemented in quantum information theory [99]. The term "entanglement" was introduced by Schrödinger [100] when he analyzed the so-called Einstein–Podolsky–Rosen paradox [101], see also discussions in Refs. [102–104].

* *
*

All experimental data known to date agree with the Pauli exclusion principle. Some theoretical ideas and experimental searches for possible violations of the Pauli exclusion principle were discussed by Okun [105] and recently by Ignatiev [37]; the published experimental tests of the validity of the Pauli exclusion principle were classified in the review report by Gillaspy [106]. Below I discuss a widespread spectroscopic approach.

The systematic spectroscopic study of the validity of the Pauli exclusion principle for electrons has been recently carried out by the Violation Pauli (VIP) collaboration. In their experiments they performed a search of X-rays from the Pauli-forbidden atomic transition from the $2p$ shell to the closed $1s^2$ shell of Cu atoms, forming the non-Pauli $1s^3$ shell, see Fig. 1.2.[9] The obtained probability that the Pauli exclusion principle is violated, according to their last measurements [108, 109], was

$$\frac{1}{2}\beta^2 < 6 \times 10^{-29}. \tag{1.14}$$

[9] It should be mentioned that it is a quite widespread approach, which at present is usually used in the experimental verification of the Pauli exclusion principle after the Ramberg and Snow experiment [107].

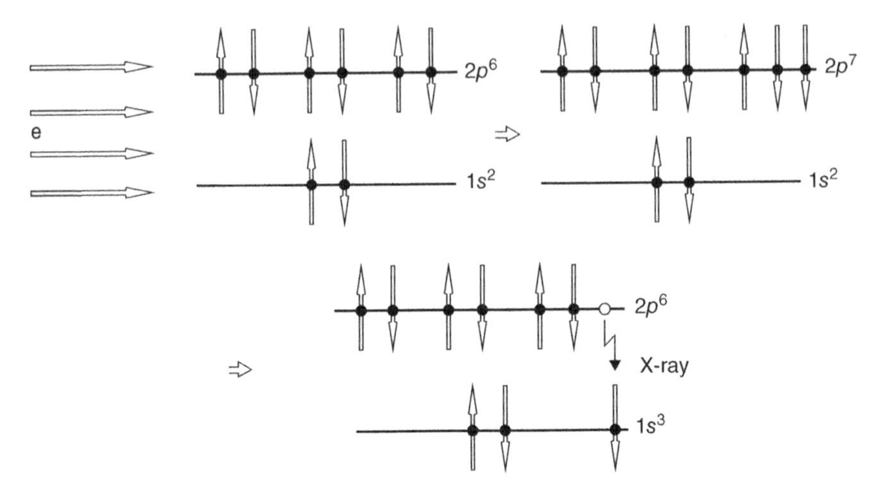

Fig. 1.2 The schematic representation of the formation of the Pauli-forbidden atomic inner-shell populations in the experimental search of the non-Pauli electrons

In the experiments performed in the Los Alamos laboratory by Elliott et al. [110] Pb instead of Cu was used. They reported a much stronger limit on the violation of the Pauli exclusion principle for electrons. Namely:

$$\frac{1}{2}\beta^2 < 2.6 \times 10^{-39}. \tag{1.15}$$

It must be mentioned that this limit was obtained by a modified method of the processing of the experimental data. As noted in Ref. [110], in the conductor there are two kinds of electrons: the current electrons that have no previous contacts with the target and the electrons within the target, which are "less new." The authors [110] took into account *all* free electrons. The application of this approach to the VIP data also changes their limit on 10 orders. On the other hand, it seems that the processing method used in Ref. [110] cannot be rigorously based. In any case, as follows from experimental data, the probability of formation of the non-Pauli states is practically zero.

It is worthwhile to make several comments in connection with the described above experimental verifications of the Pauli exclusion principle. First, usually experimenters consider the violation of the Pauli exclusion principle as a small admixture of the symmetric wave functions to the antisymmetric ones. They start from the Ignatiev–Kuzmin [111] and Greenberg–Mohapatra [112] theoretical suggestions, last years from the quon theory [113]. These theories are based on the second quantization formalism, in which only the symmetric and antisymmetric states are defined. In general, this limitation on the permutation symmetry of the possible states is not valid, because the solutions of the Schrödinger equation may belong to any representation of the permutation group, see below Eq. (1.16) and its discussion. If the Pauli exclusion principle is violated, it means that there are some electrons described by wave functions with an arbitrary permutation symmetry, not necessarily the symmetric one, see also discussion in Chapter 3.

The electrons not satisfying the Pauli exclusion principle are not described by the antisymmetric wave functions; therefore, they may not be mixed with the normal electrons that are the fermions. The transitions between states with different permutation symmetry are strictly forbidden (superselection rule). Thus, the transitions may take place only inside this group of probable non-Pauli electrons and these electrons are not identical to the "normal" Pauli electrons; in the other case they must be characterized by the antisymmetric wave functions, compare in this connection the comments by Amado and Primakoff [114]. It is also important to stress that every system of identical particles is characterized by one of the irreducible representations of the permutation group, but not by its superposition. If particles are characterized by some superposition of irreducible representations of the permutation group, they are not satisfied by the indistinguishability principle and are not identical; whereas the Pauli exclusion principle was formulated for systems of identical particles.

And last but not the least: Since the Pauli and probable non-Pauli electrons may not possess the same permutation symmetries, it is quite doubtful that the non-Pauli electrons can be located on the filled fermionic shells. We must take into account that the energy of identical particle system depends upon its permutation symmetry. The energy level separation for non-Pauli electrons can be in another energy region than measured X-ray transitions in experiments. Thus, even if really a small part of electrons exists that does not obey the Pauli exclusion principle, these non-Pauli electrons cannot be detected in the experiments described above.

The history of creation of the Pauli exclusion principle and consequent studies clearly indicate that it was not derived from the concepts of quantum mechanics, but was based on the analysis of experimental data. Pauli himself was never satisfied by this. In his Nobel Prize lecture Pauli said [5]:

> Already in my initial paper, I especially emphasized the fact that I could not find a logical substantiation for the exclusion principle nor derive it from more general assumptions. I always had a feeling, which remains until this day, that this is the fault of some flaw in the theory.

Let us stress that this was said in 1946, or after the famous Pauli theorem [115] of the relation between spin and statistics. In this theorem, Pauli did not give a direct proof. He showed that due to some physical contradictions, the second quantization operators for particles with integral spins cannot obey the fermion commutation relations; while for particles with half-integral spins their second quantization operators cannot obey the boson commutation relations. Thus, according to the

Pauli theorem, the connection between the value of spin and the permutation symmetry of many-particle wave function, Eq. (1.1), follows if we assume that particles can obey only two types of commutation relations: boson or fermion relations. At that time it was believed that it is really so. However, Pauli was not satisfied by such kind of negative proof. Very soon it became clear that he was right.

In 1953, Green [116] (then independently Volkov [117]) showed that more general paraboson and parafermion trilinear commutation relations, satisfying all physical requirements and containing the boson and fermion commutation relations as particular cases, can be introduced. A corresponding parastatistics is classified by its rank p. For the parafermi statistics p is the maximum occupation number. For $p = 1$ the parafermi statistics becomes identical to the Fermi–Dirac statistics. For the parabose statistics there are no restrictions in the occupation numbers; for $p = 1$ the parabose statistics is reduced to the Bose–Einstein statistics (for more details, see the book by Ohnuki and Kamefuchi [118] and Chapter 5).

So far the elementary particles obeying the parastatistics are not detected. Although, as discussed in Refs. [119–121], the ordinary fermions, which differ by some intrinsic properties (e.g., charge or color), but are similar dynamically, can be described by the parafermi statistics. In this case, fermions with *different* internal quantum numbers are considered as *different*, or as different states of dynamically equivalent particles. So, *quarks* with three colors obey the parafermi statistics of rank $p = 3$; *nucleons* in nuclei (isotope spin 1/2) obey the parafermi statistics of rank $p = 2$. It is important to stress that the parafermi statistics of rank p describes systems with p different types of fermions. The total wave function for such parafermions always can be constructed as an antisymmetric function in full accordance with the Pauli exclusion principle.

In 1976, Kaplan [122] revealed that the parafermi statistics is realized for quasiparticles in a crystal lattice, for example, the Frenkel excitons or magnons, but due to a periodical crystal field, the Green trilinear commutation relations are modified by the quasi-impulse conservation law, see also Ref. [123]. Later on, it was shown that the modified parafermi statistics [122] introduced by Kaplan is valid for different types of quasiparticles in a periodical lattice: polaritons [124, 125], defectons [126], the Wannier–Mott excitons [127], delocalized holes in crystals [128], delocalized coupled hole pairs [129], and some others, see also Ref. [130].

The study of properties of the quasiparticles in a periodical lattice revealed [122, 128] that even in the absence of dynamical interactions, the quasiparticle system is characterized by some kinematic interaction depending on the deviation of their statistics from the Bose (Fermi) statistics. This kinematic interaction mixes all states of the quasiparticle band. One cannot define an independent quasiparticle in a definite state. The ideal gas of such quasiparticles does not exist fundamentally. There are also no direct connection between the commutation relations for quasiparticle operators and the permutation symmetry of many-quasiparticle wave functions.

Let us return to the Pauli theorem [115]. Since there are no prohibitions on the existence of elementary particles obeying the parastatistics commutation relations, the proof [115] loses its base. After 1940, numerous proofs of the spin–statistics connection were published; see, for instance, Refs. [131–134] and the Pauli criticism [135] on the Feynman [136] and Schwinger [137] approaches. In the comprehensive book by Duck and Sudarshan [138] and in their review [139] they criticized all proofs of the spin–statistics connection published at that time except the Sudarshan proof [140]. Only this proof they assumed as correct and elementary understandable. However, in his critical review on the Duck–Sudarshan book [138], Wightman [141] noted that none of the authors criticized in the Duck–Sudarshan book will find the proof [140] satisfactory.

In his famous lectures [142] Feynman asked:

> Why is it that particles with half-integral spin are Fermi particles whose amplitudes add with the minus sign, whereas particles with integral spin are Bose particles whose amplitudes add with the positive sign? We apologize for the fact that we cannot give you an elementary explanation. An explanation has been worked out by Pauli from complicated arguments of quantum field theory and relativity. He has shown that two must necessarily go together, but we have not been able to find a way of reproducing his arguments on an elementary level. It appears to be one of the few places in physics where there is a rule which can be stated very simply, but for which no one found a simple and easy explanation. The explanation is deep down in relativistic quantum mechanics. This probably means that we do not have a complete understanding of the fundamental principle involve.

Unfortunately, after that time there has not been any progress in this direction. Most proofs of the spin–statistics connection contain negative statements; they demonstrate that abnormal cases cannot exist, but they did not answer why the normal case exists. We still have no answer what are the physical reasons that identical particles with half-integer spin are described by antisymmetric functions and identical particles with integer spin are described by completely symmetric functions. As Berry and Robbins [143] emphasized, the relation between spin and statistics "cries out for understanding."[10]

It is worthwhile to stress that the Pauli exclusion principle is not reduced only to the spin–statistics connection. It can be considered from the point of the restrictions on the allowed symmetry types of many-particle wave functions. Namely, only two types of permutation symmetry are allowed: symmetric and antisymmetric. Both

---

[10] After this Feynman comment, many studies were published, in which authors claimed that they fulfilled the Feynman requirement and proposed a simple explanation of the spin–statistics connection. However, none of these proofs can be considered rigorous, including recent publications by Jabs [144] and based on it a relativistic proof by Bennett [145].

belong to the one-dimensional representations of the permutation group; all other types of permutation symmetry are forbidden. However, the Schrödinger equation is invariant under any permutation of identical particles. The Hamiltonian of an identical particle system commutes with the permutation operators,

$$[P,H]_{-} = 0. \tag{1.16}$$

From this it follows that the solutions of the Schrödinger equation may belong to any representation of the permutation group, including multidimensional representations.

The following question might be asked:

whether this limitation on the solutions of the Schrödinger equation follows from the fundamental principles of quantum mechanics or it is an independent principle?

In Chapter 3 I discuss possible answers to this question. Here I only would like to mention that in my publications [146–148] it was rigorously proved that the indistinguishability principle is insensitive to the permutation symmetry of the wave function and cannot be used as a criterion for the verification of the Pauli exclusion principle. All published up-to-date proofs that only two types of the permutation symmetry can exist are incorrect, including the proof in the famous book by Landau and Lifshitz *Quantum Mechanics*, translated in many languages, see section 61 in English translation [149] of this book.

Thus, quantum mechanics allows all types of the permutation symmetry. In this aspect in a rigorous proof of the spin–statistics connection the possibility of the multidimensional representations of the permutation group must also be considered. However, the second quantization formalism is developed only for the symmetric and antisymmetric representations. This makes the proof of the spin–statistics connection even more improbable.

It should be mentioned that in our studies [146–148] of different scenarios following from the allowance of multidimensional representations of the permutation group, it was demonstrated that the latter leads to some contradictions with the concepts of particle identity and their independency from each other, see discussion in Section 3.2. Thus, the existence in Nature of only the one-dimensional permutation symmetries is not occasional.

It seems that at this point it makes sense to conclude this historical survey, because many of problems mentioned above will be discussed in detail in the following chapters of this book. As we saw, the Pauli exclusion principle plays the decisive role in a very wide range of phenomena: from the structure of nuclei, atoms, molecules, and solids to the formation of stars, for instance, the white dwarfs.

References

[1] W. Pauli, *Science* **103**, 213 (1946).
[2] W. Pauli, *Z. Phys.* **31**, 373 (1925).
[3] E.C. Stoner, *Philos. Mag.* **48**, 719 (1924).
[4] W. Pauli, *Z. Phys.* **31**, 765 (1925).
[5] W. Pauli, Exclusion principle and quantum mechanics, in *Nobel Lectures, Physics, 1942–1962*, Elsevier, Amsterdam, 1964, pp. 27–43.
[6] R. Kronig, The turning point, in *Theoretical Physics in the Twentieth Century*, eds. M. Fierz and V.F Weisskopf, Cambridge University Press, Cambridge, Massachusetts, 1960, pp. 5–39.
[7] B.L. van der Waerden, Exclusion principle and spin, in *Theoretical Physics in the Twentieth Century*, eds. M. Fierz and V.F Weisskopf, Cambridge University Press, Cambridge, Massachusetts, 1960, pp. 199–244.
[8] G.E. Uhlenbeck, *Oude en Nieuwe Vragen der Natuurkund*, North-Holland Publishing Co., Amsterdam, 1955.
[9] G.E. Uhlenbeck and S. Goudsmit, *Naturwissenschaften* **13**, 954 (1925).
[10] G.E. Uhlenbeck and S. Goudsmit, *Nature* **117**, 264 (1926).
[11] L.H. Thomas, *Nature* **117**, 514 (1926).
[12] W. Heisenberg, *Z. Phys.* **38**, 411 (1926).
[13] P.A.M. Dirac, *Proc. Roy. Soc. Lond. A* **112**, 621 (1926).
[14] W. Heisenberg, *Z. Phys.* **39**, 499 (1926).
[15] J.C. Slater, *Phys. Rev.* **34**, 1293 (1929).
[16] E. Fermi, *Z. Phys.* **36**, 902 (1926).
[17] S.N. Bose, *Z. Phys.* **26**, 178; 27, 384 (1924).
[18] A. Einstein, *Berl. Ber.* 261 (1924).
[19] A. Einstein, *Berl. Ber.* **1**, 18 (1925).
[20] W. Heisenberg and P. Jordan, *Z. Phys.* **37**, 263 (1926).
[21] W. Pauli, *Z. Phys.* **43**, 601 (1927).
[22] P.A.M. Dirac, *Proc. Roy. Soc. Lond. A* **117**, 610; 118, 351 (1928).
[23] B.L. van der Waerden, *Nachr. Ges. Wiss. Göttingen Math.-Phys.* 100 (1929).
[24] J. von Neumann, *Math. Z.* **30**, 3 (1927).
[25] E. Wigner, *Z. Phys.* **43**, 624 (1927).
[26] J. von Neumann and E. Wigner, *Z. Phys.* **47**, 203 (1928).
[27] J. von Neumann and E. Wigner, *Z. Phys.* **49**, 73; 51, 844 (1928).
[28] H. Weyl, *Gruppentheorie und Quantenmechanik*, Hirzel Ferlag, Leipzig, 1928.
[29] E. Wigner, *Gruppentheorie und ihre Anwendung auf die Quantenmechanik der Atomspectren*, Vieweg, Braunschweig, 1931.
[30] B.L. van der Waerden, *Die Gruppentheoretische Methode in der Quantenmechanik*, Springer, Berlin, 1932.
[31] E. Wigner, *Math. Naturwiss. Anz. Ung. Akad. Wiss.* **46**, 576 (1929).
[32] P. Ehrenfest and J.R. Oppenheimer, *Phys. Rev.* **37**, 333 (1931).
[33] M. Born and J.R. Oppenheimer, *Ann. Phys. (Leipzig)* **84**, 457 (1927).
[34] H.G. Dieke and H.D. Babcock, *Proc. Natl. Acad. Sci. U. S. A.* **13**, 670 (1927).
[35] P.A.M. Dirac, *Proc. Roy. Soc. Lond. A* **114**, 243 (1927).
[36] P. Jordan and E. Wigner, *Z. Phys.* **47**, 631 (1928).
[37] A.Yu. Ignatiev, *Radiat. Phys. Chem.* **75**, 2090 (2006).
[38] J. Chadwick, *Nature* **129**, 312 (1932).
[39] W. Heisenberg, *Z. Phys.* **77**, 1 (1932).
[40] E. Wigner, *Phys. Rev.* **51**, 106 (1937).
[41] F. Hund, *Z. Phys.* **105**, 202 (1937).
[42] E. Feenberg and E. Wigner, *Phys. Rev.* **51**, 95 (1937).
[43] G. Racah, *Phys. Rev.* **61**, 186 (1942).

[44] G. Racah, *Phys. Rev.* **62**, 438 (1942).
[45] G. Racah, *Phys. Rev.* **63**, 367 (1943).
[46] G. Racah, *Phys. Rev.* **76**, 1352 (1949).
[47] E.U. Condon and G.H. Shortley, *The Theory of Atomic Spectra*, Cambridge University Press, New York, 1935.
[48] H.A. Jahn, *Proc. Roy. Soc. Lond. A* **201**, 516 (1950).
[49] H.A. Jahn, *Proc. Roy. Soc. Lond. A* **205**, 192 (1951).
[50] H.A. Jahn and H. van Wieringen, *Proc. Roy. Soc. Lond. A* **209**, 502 (1951).
[51] J.P. Elliott, J. Hope, and H.A. Jahn, *Philos. Trans. Roy. Soc. Lond. A* **246**, 241 (1953).
[52] H.A. Jahn, *Phys. Rev.* **96**, 989 (1954).
[53] B.H. Flowers, *Proc. Roy. Soc. Lond. A* **212**, 248, 1952
[54] A.R. Edmonds and B.H. Flowers, *Proc. Roy. Soc. Lond. A* **214**, 515 (1952).
[55] I.G. Kaplan, *Sov. Phys. JETP* **10**, 747 (1960).
[56] I.G. Kaplan, *Sov. Phys. JETP* **24**, 114 (1967).
[57] I.G. Kaplan and O.B. Rodimova, *Int. J. Quantum Chem.* **10**, 699 (1976).
[58] I.G. Kaplan, *Sov. Phys. JETP* **14**, 401 (1962).
[59] P. Kramer, *Z. Phys.* **205**, 181 (1967).
[60] P. Kramer, *Z. Phys.* **216**, 68 (1968).
[61] V.G. Neudachin, Yu.F. Smirnov, and N.F. Golovanova, *Adv. Nucl. Phys.* **11**, 1 (1979).
[62] I.G. Kaplan, *Sov. Phys. JETP* **14**, 568 (1962).
[63] M. Kotani and M. Siga, *Proc. Phys. Math. Soc. Jpn.* **19**, 471 (1937).
[64] M. Kotani, A. Amemiya, E. Ishiguro, and T. Kimura, *Tables of Molecular Integrals*, Maruzen, Tokyo, 1955.
[65] M. Kotani, Y. Mizuno, K. Kayama, and E. Ishiguro, *Proc. Phys. Soc. Jpn.* **12**, 707 (1957).
[66] I.G. Kaplan, *Sov. Phys. JETP* **17**, 261 (1963).
[67] I.G. Kaplan, *Liet. Fiz. Rinkinys* **3**, 227 (1963).
[68] I.G. Kaplan, *Theor. Exp. Chem.* **1**, 399, 407 (1965); 2, 335 (1966); 3, 83 (1967).
[69] I.G. Kaplan, *Symmetry of Many-Electron Systems*, Academic Press, New York, 1975 (translation by J. Gerratt of enlarged Russian Edition, Nauka, Moscow, 1969).
[70] I.G. Kaplan and O.B. Rodimova, *Int. J. Quantum Chem.* **7**, 1203 (1973).
[71] F.A. Matsen, *Adv. Quantum Chem.* **1**, 60 (1964).
[72] F.A. Matsen, *J. Phys. Chem.* **68**, 3282 (1964).
[73] F.A. Matsen, A.A. Cantu, and R.D. Poshusta, *J. Phys. Chem.* **70**, 1558 (1966).
[74] F.A. Matsen, *J. Am. Chem. Soc.* **92**, 3525 (1970).
[75] W.A. Goddard III, *Phys. Rev.* **73, 81**, 157 (1967).
[76] W.A. Goddard III, *J. Chem. Phys.* **48**, 450 (1968).
[77] G.A. Gallup, *J. Chem. Phys.* **48**, 1752 (1968).
[78] G.A. Gallup, *J. Chem. Phys.* **50**, 1206 (1969).
[79] J. Gerratt, *Annu. Rep. Chem. Soc.* **65**, 3 (1968).
[80] J. Gerratt, *Adv. At. Mol. Phys.* **7**, 141 (1971).
[81] M.D. Girardeau, *J. Math. Phys.* **4**, 1096 (1963).
[82] D. Girardeau, *Phys. Rev. Lett.* **27**, 1416 (1971).
[83] J.D. Gilbert, *J. Math. Phys.* **18**, 791 (1977).
[84] C.K. Law, *Phys. Rev. A* **71**, 034306 (2005).
[85] P. Sancho, *J. Phys. A Math. Gen.* **39**, 12525 (2006).
[86] J. Bardeen, L.N. Cooper, and J.R. Schrieffer, *Phys. Rev.* **106**, 162 (1957); 108, 1175 (1957).
[87] L.N. Cooper, *Phys. Rev.* **104**, 1189 (1956).
[88] J.R. Schrieffer, *Theory of Superconductivity*, Addison-Wesley, Redwood City, California, 1988.
[89] I.G. Kaplan, O. Navarro, and J.A. Sanchez, *Physica C* **419**, 13 (2005).
[90] I.G. Kaplan and O. Navarro, Comparative study of statistics of Cooper's electron pairs and coupled holes pairs in high-T_c ceramics, in *New Topic in Superconductivity Research*, ed. B.P. Martins, Nova Science, New York, 2006, pp. 223–237.

[91] Ya.B. Zeldovich, *Sov. Phys. JETP* **10**, 403 (1960).
[92] S. Huzinaga, D. McWilliams, and A.A. Cantu, *Adv. Quantum Chem.* **7**, 187 (1973).
[93] J.L. Pascual, N. Barros, Z. Barandiarán, and L. Seijo, *J. Phys. Chem. A* **113**, 12454 (2009).
[94] G. Gutowski and L. Piela, *Mol. Phys.* **64**, 337 (1988).
[95] Ł. Rajchel, P.S. Zuchowski, M.M. Szcześniak, and G. Chałasiński, *Chem. Phys. Lett.* **486**, 160 (2010).
[96] R.H. Fowler, *Mon. Not. R. Astron. Soc.* **87**, 114 (1926).
[97] D. Cavalcanti, L.M. Malard, F.M. Matinaga, M.O. Terra Cunha, and M. França Santos, *Phys. Rev. B* **76**, 113304 (2007).
[98] R. Horodecki, P. Horodecki, M. Horodecki, and K. Horodecki, *Rev. Mod. Phys.* **81**, 867 (2009).
[99] M.A. Nielsen and I.L. Chuang, *Quantum Computation and Quantum Information*, Cambridge University Press, Cambridge, Massachusetts, 2000.
[100] E. Schrödinger, *Math. Proc. Camb. Philos. Soc.* **31**, 555 (1935).
[101] A. Einstein, B. Podolsky, and N. Rosen. *Phys. Rev.* **47**, 777 (1935).
[102] J.S. Bell, *Speakable and Unspeakable in Quantum Mechanics*, Cambridge University Press, Cambridge, Massachusetts, 2000.
[103] D.M. Greenberger, M.A. Horne, A. Shimony, and A. Zeilinger, *Am. J. Phys.* **58**, 1131 (1990).
[104] N.D. Mermin, *Am. J. Phys.* **58**, 731 (1990).
[105] L.B. Okun, *Sov. Phys. Usp.* **32**, 543 (1989).
[106] J.D. Gillaspy, High precision experimental tests of the symmetrization postulate for fermions, in *Spin-Statistics Connection and Commutation Relations*, eds. R.C. Hilborn and G.M. Tino, AIP Conf. Proc., 2000, vol. **545**, pp. 241–252.
[107] E. Ramberg and G.A. Snow, *Phys. Lett. B* **238**, 439 (1990).
[108] C. Circeanu (Petrascu) et al. *Found. Phys.* **41**, 282 (2011).
[109] C. Curceanu (Petrascu) et al. *Int. J. Quantum Inf.* **9**, 145 (2011).
[110] S.R. Elliott, B.H. LaRoque, V.M. Gehman, M.F. Kidd and M. Chen, *Found. Phys.* **42**, 1015 (2012).
[111] A.Yu. Ignatiev and V.A. Kuzmin, *Sov. J. Nucl. Phys.* **461**, 786 (1987).
[112] O.W. Greenberg and R.N. Mohapatra, *Phys. Rev. Lett.* **59**, 2507 (1987).
[113] O.W. Greenberg, *Phys. Rev. Lett.* **64**, 705 (1990).
[114] R.D. Amado and H. Primakoff, *Phys. Rev. C* **22**, 1338 (1980).
[115] W. Pauli, *Phys. Rev.* **58**, 716 (1940).
[116] H.S. Green. *Phys. Rev.* **90**, 270 (1953).
[117] D.V. Volkov. *Sov. Phys. JETP* **9**, 1107 (1959).
[118] Y. Ohnuki and S. Kamefuchi, *Quantum Field Theory and Parastatistics*, Springer, Berlin, 1982.
[119] N.A. Chernikov, *Acta Phys. Pol.* **21**, 52 (1962).
[120] O.W. Greenberg, *Phys. Rev. Lett.* **13**, 598 (1964).
[121] A.B. Govorkov. *Sov. J. Part. Nucl.* **14**, 520 (1983).
[122] I.G. Kaplan, *Theor. Math. Phys.* **27**, 254 (1976).
[123] I.G. Kaplan and M.A. Ruvinskii, *Sov. Phys. JETP* **71**, 2142, 1976.
[124] A.N. Avdyugin, Yu.D Zavorotnev, and L.N. Ovander. *Sov. Phys. Solid State* **25**, 1437 (1983).
[125] B.A. Nguyen, *J. Phys. C Solid State Phys.* **21**, L1209 (1988).
[126] D.I. Pushkarov. *Phys. Status Solidi B* **133**, 525 (1986).
[127] A. Nguyen and N.C. Hoang. *J. Phys. Condens. Matter* **2**, 4127 (1990).
[128] I.G. Kaplan and O. Navarro, *J. Phys. Condens. Matter* **11**, 6187 (1999).
[129] I.G. Kaplan and O. Navarro, *Physica C* **341**(348), 217 (2000).
[130] I.G. Kaplan, The Pauli exclusion principle, spin-statistics connection, and permutation symmetry of many-particle wave functions, in *Fundamental World of Quantum Chemistry*, eds. E.J. Brandas and E.S. Kryachko, Kluwer, Dordrecht, 2003, vol. **1**, pp. 183–220.
[131] D. Hall, A.S. Wightman, and K. Dan, *Vidensk. Selsk. Mat. Fys. Medd.* 31, (1957).
[132] G. Luders and B. Zumino, *Phys. Rev.* **110**, 1450 (1958).
[133] N. Burgoyne, *Nuovo Cimento* **8**, 607 (1958).

[134] A.A. Broyles, *Am. J. Phys.* **44**, 340 (1976).
[135] W. Pauli, *Prog. Theor. Phys.* **5**, 526 (1950).
[136] R.P. Feynman, *Phys. Rev.* **76**, 749 (1949).
[137] J. Schwinger, *Phys. Rev.* **74**, 1939 (1948); 82, 914 (1951).
[138] I. Duck and E.C.G. Sudarshan, *Pauli and the Spin-Statistics Theorem*, World Scientific, Singapore, 1997.
[139] I. Duck and E.C.G. Sudarshan, *Am. J. Phys.* **66**, 284 (1998).
[140] E.C.G. Sudarshan, *Stat. Phys. Suppl. J. Indian. Inst. Sci.* **June**, 123 (1975).
[141] A.S. Wightman, *Am. J. Phys.* **67**, 742 (1999).
[142] In *Feynman Lectures on Physics*, R.P. Feynman, R.B. Leighton, and M. Sands, Addison-Wesley, Reading, Massachusetts, 1965, vol. **III**, p.3.
[143] M. Berry and J. Robbins, Quantum indistinguishability: spin-statistics without relativity or field theory? in *Spin-Statistics Connection and Commutation Relations*, eds. R.C. Hilborn and G.M. Tino, AIP Conf. Proc., 2000, vol. **545**, pp. 3–15.
[144] A. Jabs, *Found. Phys.* **40**, 776 (2010).
[145] A.F. Bennett, *Found. Phys.* **45**, 370 (2015).
[146] I.G. Kaplan, *J. Mol. Struct.* **272**, 187 (1992).
[147] I.G. Kaplan, *Int. J. Quantum Chem.* **89**, 268 (2002).
[148] I.G. Kaplan, *Found. Phys.* **43**, 1233 (2013).
[149] L.D. Landau and E.M. Lifshitz, *Quantum Mechanics, (Nonrelativistic Theory)*, 3rd edn., Pergamon Press, Oxford, 1977.

2

Construction of Functions with a Definite Permutation Symmetry

2.1 Identical Particles in Quantum Mechanics and Indistinguishability Principle

The quantum particles are characterized by two kinds of properties: *intrinsic* properties (mass, electric charge, spin, magnetic moment, etc.) and *dynamic* properties (position, momentum, and energy). The intrinsic properties are permanent for each given particle, while the dynamic properties depend on interaction forces.

> *Two particles are said to be identical if all their intrinsic properties are exactly the same.*

Thus, all electrons, protons, neutrons, nuclei for each isotope of the same element, positrons, muons, and other elementary particles are identical, since they have the same intrinsic properties.

The problem of distinguishability of identical particles is different in classical and quantum mechanics. In classical mechanics we can always enumerate (label) particles and then track its trajectories. Thus, each particle can be identified. In quantum mechanics the situation is in principle different. According to the Heisenberg uncertainty principle, the position and momentum of particle cannot be simultaneously measured precisely. If we at some moment localize two particles, their momenta will not have definite values; and in a next moment the positions of these two particles will not be defined. In quantum mechanics the concept of trajectory

The Pauli Exclusion Principle: Origin, Verifications, and Applications, First Edition. Ilya G. Kaplan.

has no sense and it is impossible to trace each particle; the identical particles lose their individuality.

It is important to mention that in quantum mechanics in some cases even not identical particles can be considered as indistinguishable. It is the case when some type of interactions can be neglected in comparison with others and (or) the difference of mass of not identical particles is small. For instance, in nuclei the electromagnetic repulsion between protons is many orders smaller that the so-called strong interaction between nucleons: protons and neutrons. As a result, the forces between all pairs of nucleons can be considered as equal (the strong interaction does not depend upon the electric charge). The kinetic terms for protons and neutrons in the Hamiltonian is almost the same, because $m_n = 1\,00\,137 m_p$. Thus, we can say that in nuclei the protons and neutrons are *similar dynamically*; they are physically indistinguishable. In addition to the ordinary spin $i = 1/2$, the nucleons are characterized by a two-valued quantum number, *isotopic* (or *isobaric*) *spin*; one value of which corresponds to proton and another value corresponds to neutron. The total wave function must be antisymmetrized in respect to permutations of all nucleons. Let us stress that the free protons and neutrons are certainly distinguishable. Some other cases when different particles become indistinguishable have been discussed by Lyuboshitz and Podgoretskii [1, 2].

The indistinguishability of identical particles in quantum mechanics causes many physical and chemical effects. I will present here two of them. In 1969, we predicted [3] the oscillation behavior in the photoelectron angular distribution from an oriented H_2 molecule. The obtained picture corresponded to the interference of two coherent waves emitted from two centers located near the nuclei of the molecule (protons in the case of H_2). Due to the indistinguishability of electrons and protons as well, one cannot say what of electrons was emitted and from what center. The probability to eject the electron is equal for each center and this produces the interference. At large energy the wave function of the emitted electron can be presented as a plane wave, $e^{i\boldsymbol{qr}}$, where $\boldsymbol{q}$ is the wave vector of the emitted electron.

If from centers a and b two plane waves are emitted, see Fig. 2.1, then in the direction θ their phase shift will be determined by the product

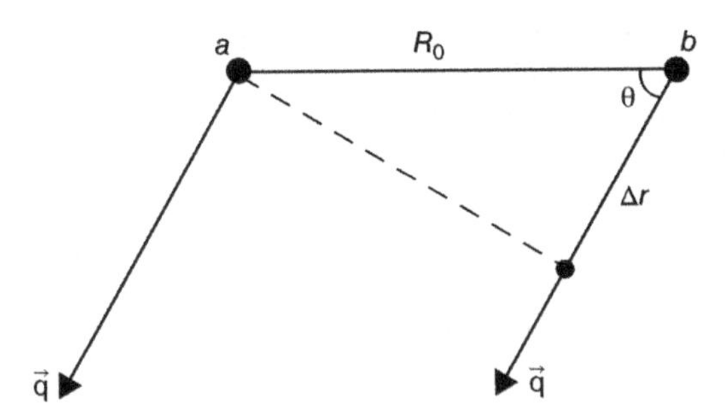

Fig. 2.1 The scheme of the photoeffect on the oriented H_2 molecule, which leads to the interference phenomena

$$q\Delta r = qR_0\cos\theta. \tag{2.1}$$

Maxima will be observed for those values of θ at which the path length Δr is an integral number of the de Broglie wave of the emitted electron

$$\lambda_e = \frac{2\pi}{q}. \tag{2.2}$$

At small photon energies the photoelectron wave length λ_e is greater than the internuclear distance R_0 and the conditions of the appearance of subsidiary maxima are not fulfilled, which explains the occurrence of the oscillatory picture only when the energy of incident photon is sufficiently large. The predicted oscillations were recently (after 38 years) confirmed in experiment [4]. It must be stressed that the described phenomenon reflects not only the quantum indistinguishability of identical particles, but also the wave properties of electrons (the particle–wave dualism).

Second example concerns the case of asymmetric dissociation limit of homoatomic dimers. As was established in experimental and theoretical studies of the Sc_2 dimer, see Ref. [5], Sc_2 in its ground state dissociates on one Sc atom in the ground state and another in an excited state. However, from the quantum indistinguishability of the identical atoms it follows that each of the Sc atoms can be excited and it is impossible to determine which atom was excited. The wave function of dimer in the dissociation limit must be presented as the linear combinations of the product of atomic wave functions corresponding to the case when one Sc atom is in the ground state and another Sc atom is in an excited state, and the product of wave functions describing the opposite possibility:

$$\Psi(Sc_2, R\to\infty) = \frac{1}{\sqrt{2}}\left[\Psi_0(Sc_a)\Psi_n^*(Sc_b) \pm \Psi_n^*(Sc_a)\Psi_0(Sc_b)\right]. \tag{2.3}$$

Thus, although the atoms dissociate in different states, both atoms have the same electronic orbital population. On the other hand, from Eq. (2.3) it follows that in the case of the dissociation of identical atoms in different states, the even (g) and odd (u) terms in the spectroscopic classification can be constructed. This corresponds to the accepted classification of the dimer terms when it dissociates on the atoms in different states [6–8]. The group-theoretical methods of classification of the Pauli allowed states for molecules and for atoms as well we discuss in detail in Chapter 4.

The indistinguishability of identical particles in quantum mechanics is the reason that *the permutation group is the symmetry group of an identical particle system.* It means that the quantum states of an identical particle system are classified according to the irreducible representations of the permutation group (the necessary information about the permutation group see in Appendix B).

The solutions of the Schrödinger equation for the identical particle system must belong to one of the irreducible representations of the permutation group. This provides the possibility of formulating the Pauli exclusion principle via the permutation symmetry of many-particle wave functions. For the system of identical fermions,

$$P_{ij}\Psi(1,\ldots,i,\ldots j,\ldots,N) = -\Psi(1,\ldots,i,\ldots j,\ldots,N) \tag{2.4}$$

for each transposition. The fermion wave functions belong to the antisymmetric irreducible representation, $\Gamma^{[1^N]}$, of the permutation group π_N, while the boson wave functions belong to the symmetric irreducible representation, $\Gamma^{[N]}$, of the same group and are completely symmetric, as it follows from the general formulation of the Pauli exclusion principle represented in Chapter 1, p. 9.

Strictly speaking, wave functions of all identical microparticles must be symmetrized. However, in reality it is not so. At large interparticle separation, the difference between the energy (and other properties) of identical particle system described by antisymmetric (symmetric) wave functions and completely nonsymmetrical wave functions is negligible. In Section 2.3.1 we consider the two-electron system and show that the antisymmetrization leads to a so-called *exchange energy*, which exponentially decreases with distance. The influence of the antisymmetrization on the properties of two electrons, one is on the Earth and another is on the Moon [9], Mars [10], or even in another galaxy [11], is the same, as in the case of two remote electrons in one laboratory sample. Just this allows studying isolated atoms or molecules.

However, it is important to mention that the situation is different in the case if we study the statistical properties of some system. The point is that in quantum gases the statistics are determined by the occupation numbers of one-particle states which follow from the Pauli exclusion principle. As a result, the properties of Bose- and Fermi-gas are different even in the absence of any interactions, including the exchange interactions. The total wave function must be antisymmetrized or symmetrized for all particles in the considered gas volume, providing the right statistics, and this must be performed even in the case of the ideal gas.

The indistinguishability of identical particles means that the properties of any N identical particle system must be invariant under arbitrary permutation of these particles. In quantum mechanics the physical properties of a system in a quantum state n is determined by the probability density of this state. If the state is described by the definite wave function $\Psi_n(1, \ldots, N)$, its probability density is proportional to $|\Psi_n(1, \ldots, N)|^2$. The mathematical formulation of the indistinguishability principle can be presented as

$$P|\Psi_n(1,\ldots,N)|^2 = |\Psi_n(1,\ldots,N)|^2, \tag{2.5}$$

where P is any permutation of the group π_N. It is important to mention that only the nondegenerate states can be the pure states, that is, the states which are characterized by a wave function. In a degenerate permutation state we cannot select a pure state, as it can be done in the case of the point groups, since for this we should delete degeneracy. However, the latter cannot be achieved without violating the particle identity with respect to permutations. The degenerate permutation states belong to so-called mixed states. We discuss the case of the degenerate states in Section 3.1.

2.2 Construction of Permutation-Symmetric Functions Using the Young Operators

The wave function of an arbitrary identical particle system must satisfy the Pauli exclusion principle, that is, be symmetric for bosons and antisymmetric for fermions. On the other hand, for Hamiltonian, in which relativistic interactions are neglected (they can be subsequently taken into account as a perturbation), the total spin S of the system is a good quantum number and the wave function must be the eigenfunction of the operator $\hat{S}^2$. In this section, we will describe how to construct the many-particle wave function with definite permutation symmetry. The construction of wave functions describing states with the total spin and simultaneously satisfying the Pauli exclusion principle is discussed in next section of this chapter.

The permutation symmetry is classified by the so-called *Young diagram*, each corresponding to some irreducible representation of the permutation group, see Appendix B. The Young diagrams are denoted by the letter λ in brackets and depicted as graphs.

$$[\lambda] = [\lambda_1 \lambda_2 \ldots \lambda_k],$$
$$\lambda_1 \geq \lambda_2 \geq \cdots \geq \lambda_k, \quad \sum_{i=1}^{k} \lambda_i = N. \tag{2.6}$$

Here, λ_i is represented by a row of λ_i cells. The presence of several rows of identical length λ_i is indicated by a power of λ_i. We represent below the Young diagrams for groups $\pi_2 - \pi_4$.

For two cells one can form only two Young diagrams:

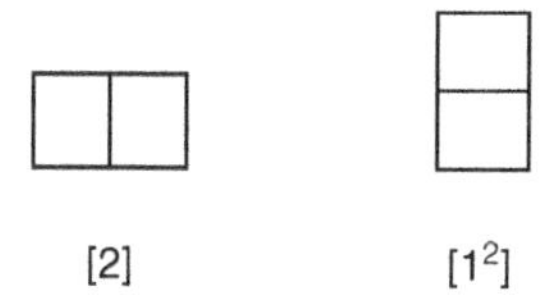

For the permutation group of three elements, π_3, one can form from three cells three Young diagrams:

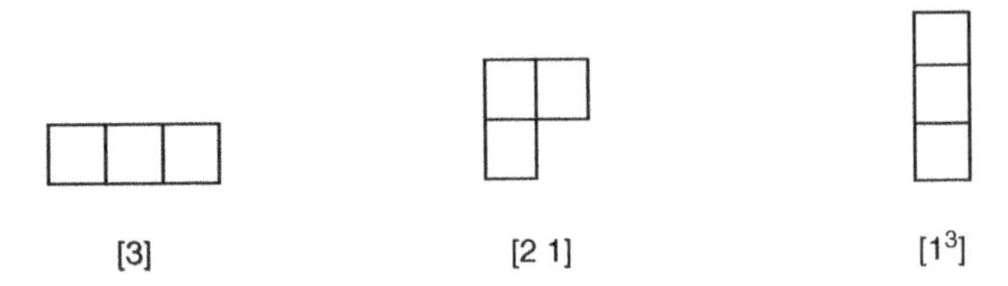

The group π_4 has five Young diagrams:

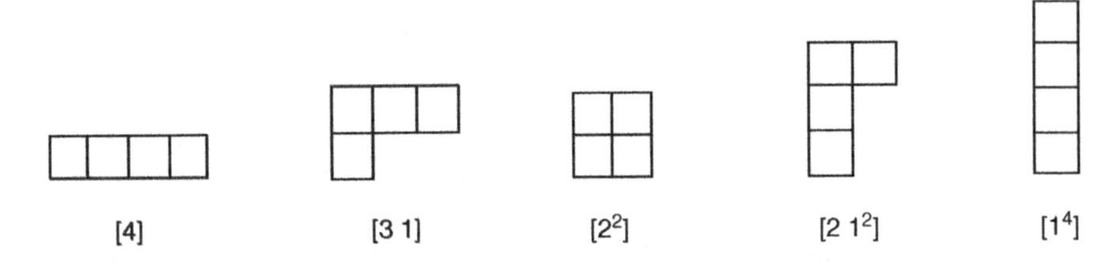

Each Young diagram $[\lambda_N]$ uniquely corresponds to a specific irreducible representation $\Gamma^{[\lambda_N]}$ of the group π_N. The assignment of a Young diagram determines the permutation symmetry of the basis functions for an irreducible representation, that is, determines the behavior of the basis functions under permutations of their arguments. A diagram with only one row corresponds to a function symmetric in all its arguments. A Young diagram with one column corresponds to a completely antisymmetric function. All other types of diagrams correspond to intermediate types of symmetry. There are certain rules that enable one to find the matrices of irreducible representations of the permutations group from the form of the corresponding Young diagram. Such rules are especially simple in the case of the so-called standard orthogonal representation, or the *Young–Yamanouchi representation*; see Appendix B, Section B.2.

The basis functions for an irreducible representation $\Gamma^{[\lambda]}$ can be constructed by means of the so-called normalized *Young operators*, Eq. (B.33),[1]

$$\omega_{rt}^{[\lambda]} = \sqrt{\frac{f_\lambda}{N!}} \sum_P \Gamma_{rt}^{[\lambda]}(P) P, \tag{2.7}$$

where the summation over P runs over all the $N!$ permutations of the group π_N, $\Gamma_{rt}^{[\lambda]}(P)$ are the matrix elements of the Young–Yamanouchi orthogonal

[1] Operator (2.7) should not be mixed up with the operator that symmetrizes the rows and antisymmetrizes the columns in Young diagram, which is also named as the Young operator; see Ref. [12].

representation, and f_λ is the dimension of the irreducible representation $\Gamma^{[\lambda]}$. The application of operator (2.7) to a nonsymmetrized product of orthonormal one-particle functions φ_a

$$\Phi_0 = \varphi_1(1)\varphi_2(2)\ldots\varphi_N(N) \tag{2.8}$$

produces a normalized function

$$\Phi_{rt}^{[\lambda]} = \omega_{rt}^{[\lambda]}\Phi_0 = \sqrt{\frac{f_\lambda}{N!}}\sum_P \Gamma_{rt}^{[\lambda]}(P)P\Phi_0, \tag{2.9}$$

which transforms in accordance with the representation $\Gamma^{[\lambda]}$. Let us prove this statement.

The action of an arbitrary permutation Q of the group π_N on the function (2.9) can be presented as

$$Q\Phi_{rt}^{[\lambda]} = \sqrt{\frac{f_\lambda}{N!}}\sum_P \Gamma_{rt}^{[\lambda]}(P)QP\Phi_0 = \sqrt{\frac{f_\lambda}{N!}}\sum_R \Gamma_{rt}^{[\lambda]}(Q^{-1}R)R\Phi_0, \tag{2.10}$$

where we have denoted the permutation QP by R and made use of the invariance properties of the sum over all group elements. Further, we write the matrix element of the product of permutations as products of matrix elements and make use of property (A.29) of the orthogonal matrices:

$$\Gamma_{rt}^{[\lambda]}(Q^{-1}R) = \sum_u \Gamma_{ru}^{[\lambda]}(Q^{-1})\Gamma_{ut}^{[\lambda]}(R) = \sum_u \Gamma_{ur}^{[\lambda]}(Q)\Gamma_{ut}^{[\lambda]}(R). \tag{2.11}$$

Substituting (2.11) in (2.10), we finally obtain

$$Q\Phi_{rt}^{[\lambda]} = \sum_u \Gamma_{ur}^{[\lambda]}(Q)\Phi_{ut}^{[\lambda]}. \tag{2.12}$$

The function $\Phi_{rt}^{[\lambda]}$ transforms as the r-th column of the irreducible representation $\Gamma^{[\lambda]}$, and the set of f_λ functions $\Phi_{rt}^{[\lambda]}$ with fixed second index t forms a basis for the irreducible representation $\Gamma^{[\lambda]}$. One can form altogether f_λ independent bases corresponding to the number of different values of t. This should be expected, since $N!$ functions $P\Phi_0$ form a basis for the regular representation of π_N, and in the decomposition of the regular representation each irreducible representation occurs as many times as its dimension.

Functions belonging to the irreducible representation $\Gamma^{[\lambda]}$ can be also constructed by the permutation of the one-particle function labels in Eq. (2.8).

Let us denote such permutations by placing a bar over P. It can be easily checked that

$$\bar{P}\Phi_0 = P^{-1}\Phi_0. \tag{2.13}$$

In Appendix B it is proved, see Eq. (B.51), that

$$\bar{P}\Phi_{rt}^{[\lambda]} = \sum_u \Gamma_{ut}^{[\lambda]}(\bar{P})\Phi_{ru}^{[\lambda]}. \tag{2.14}$$

Functions $\Phi_{rt}^{[\lambda]}$ with the fixed first index transform into each other under permutations $\bar{P}$.

Thus, the first index, r, characterizes the symmetry of the function $\Phi_{rt}^{[\lambda]}$ under permutation of the arguments (electron numbers) and the second index, t, enumerating the different bases of $\Gamma^{[\lambda]}$, characterizes the symmetry of $\Phi_{rt}^{[\lambda]}$ under permutations of the labels (numbers) of one-particle functions φ_i in Eq. (2.8).

Let us write down, as an example, the basis functions (2.9) for all irreducible representations of π_3. For writing functions (2.9) in an explicit form we should know the matrix elements of the Young–Yamanouchi representation. All matrices for the Young–Yamanouchi representation for groups $\pi_2 - \pi_6$ are represented in book [13], appendix 5. For π_3, two representations are one-dimensional: $\Gamma^{[3]}(P) = 1$ for all P, while $\Gamma^{[1^3]}(P) = (-1)^p P$ (p is the parity of the permutation P). The representation $\Gamma^{[21]}$ is two-dimensional; its rows and columns are enumerated by two standard Young tableaux

$$r^{(1)} \quad \begin{array}{|c|c|}\hline 1 & 2 \\ \hline 3 & \multicolumn{1}{c}{} \\ \cline{1-1}\end{array} \qquad\qquad r^{(2)} \quad \begin{array}{|c|c|}\hline 1 & 2 \\ \hline 3 & \multicolumn{1}{c}{} \\ \cline{1-1}\end{array} \tag{2.15}$$

and is given by the following matrices:

$$\begin{array}{ccc} I & P_{12} & P_{13} \\ \begin{pmatrix} 1 & \\ & 1 \end{pmatrix} & \begin{pmatrix} 1 & \\ & -1 \end{pmatrix} & \begin{pmatrix} -\frac{1}{2} & -\frac{\sqrt{3}}{2} \\ -\frac{\sqrt{3}}{2} & \frac{1}{2} \end{pmatrix} \\ P_{23} & P_{123} & P_{132} \\ \begin{pmatrix} -\frac{1}{2} & \frac{\sqrt{3}}{2} \\ \frac{\sqrt{3}}{2} & \frac{1}{2} \end{pmatrix} & \begin{pmatrix} -\frac{1}{2} & \frac{\sqrt{3}}{2} \\ -\frac{\sqrt{3}}{2} & -\frac{1}{2} \end{pmatrix} & \begin{pmatrix} -\frac{1}{2} & -\frac{\sqrt{3}}{2} \\ \frac{\sqrt{3}}{2} & -\frac{1}{2} \end{pmatrix} \end{array}. \tag{2.16}$$

Now we can represent the basis functions for all irreducible representations of π_3. Namely:

$$\Phi^{[3]} = \frac{1}{\sqrt{6}}\{\varphi_a(1)\varphi_b(2)\varphi_c(3) + \varphi_a(2)\varphi_b(1)\varphi_c(3) + \varphi_a(3)\varphi_b(2)\varphi_c(1)$$
$$+ \varphi_a(1)\varphi_b(3)\varphi_c(2) + \varphi_a(2)\varphi_b(3)\varphi_c(1) + \varphi_a(3)\varphi_b(1)\varphi_c(2)\}.$$

$$\Phi^{[1^3]} = \frac{1}{\sqrt{6}}\{\varphi_a(1)\varphi_b(2)\varphi_c(3) - \varphi_a(2)\varphi_b(1)\varphi_c(3) - \varphi_a(3)\varphi_b(2)\varphi_c(1)$$
$$- \varphi_a(1)\varphi_b(3)\varphi_c(2) + \varphi_a(2)\varphi_b(3)\varphi_c(1) + \varphi_a(3)\varphi_b(1)\varphi_c(2)\}.$$

$$\Phi^{[21]}_{11} = \frac{1}{\sqrt{12}}\{2\varphi_a(1)\varphi_b(2)\varphi_c(3) + 2\varphi_a(2)\varphi_b(1)\varphi_c(3) - \varphi_a(3)\varphi_b(2)\varphi_c(1)$$
$$- \varphi_a(1)\varphi_b(3)\varphi_c(2) - \varphi_a(2)\varphi_b(3)\varphi_c(1) - \varphi_a(3)\varphi_b(1)\varphi_c(2)\}.$$

$$\Phi^{[21]}_{21} = \frac{1}{2}\{-\varphi_a(3)\varphi_b(2)\varphi_c(1) + \varphi_a(1)\varphi_b(3)\varphi_c(2) - \varphi_a(2)\varphi_b(3)\varphi_c(1)$$
$$+ \varphi_a(3)\varphi_b(1)\varphi_c(2)\}.$$

$$\Phi^{[21]}_{12} = \frac{1}{2}\{-\varphi_a(3)\varphi_b(2)\varphi_c(1) + \varphi_a(1)\varphi_b(3)\varphi_c(2) + \varphi_a(2)\varphi_b(3)\varphi_c(1)$$
$$- \varphi_a(3)\varphi_b(1)\varphi_c(2)\}.$$

$$\Phi^{[21]}_{22} = \frac{1}{\sqrt{12}}\{2\varphi_a(1)\varphi_b(2)\varphi_c(3) - 2\varphi_a(2)\varphi_b(1)\varphi_c(3) + \varphi_a(3)\varphi_b(2)\varphi_c(1)$$
$$+ \varphi_a(1)\varphi_b(3)\varphi_c(2) - \varphi_a(2)\varphi_b(3)\varphi_c(1) - \varphi_a(3)\varphi_b(1)\varphi_c(2)\}.$$

In the case of $N = 4$ five Young tables can be constructed. It means that the group π_4 has five irreducible representations. Each basis function of an irreducible representation $\Gamma^{[\lambda]}$ can be associated with a Young tableau, and the dimension of this representation is given by the number of standard Young tableaux, that is, by the number of different ways of distributing the N numbers in a Young diagram such that they increase left to right along the rows and down the columns. Thus, for $\Gamma^{[21^2]}$ there are three standard tableaux:

$r^{(1)}$		$r^{(2)}$		$r^{(3)}$	
1	2	1	3	1	4
3		2		2	
4		4		3	

(2.17)

In Section B.2 the simple rules for constructing the matrices of the Young–Yamanuchi representations for transpositions $P_{i-1,I}$ are presented, since all the other permutations can be expressed as a product of transpositions of the $P_{i-1,i}$ type.

The dimension of an irreducible representation may also be calculated by the following formula [14, 15]:

$$f_\lambda = \frac{N! \sum_{i<j} (h_i - h_j)}{h_1! h_2! \ldots h_m!}, \quad h_i = \lambda^{(i)} + m - i, \tag{2.18}$$

where $\lambda^{(i)}$ is the length of row i and m is the number of rows in the Young diagram $[\lambda]$. For $[\lambda] = [21^2]$ $h_1 = 2+3-1 = 4$, $h_2 = 1+3-2 = 2$, and $h_3 = 1+3-3 = 1$. According to Eq. (2.18), the dimension $f_{[21^2]} = 3$ is in full accordance with the number of standard tableaux (2.17).

We have discussed the construction of the basis functions for the arbitrary irreducible representation $\Gamma^{[\lambda]}$ of the permutation group π_N. For the construction of the wave functions of N-particle system we must take into account the spin of particle. As in the case of basis functions, we will use the so-called one-particle approximation where each particle is characterized by its own wave function. The latter can be written as a product of one-particle spatial wave function $\varphi_n(\mathbf{r})$ and spin function χ_σ,

$$\psi_{n\sigma} = \varphi_n(\mathbf{r}) \chi_\sigma, \tag{2.19}$$

where n is a set of quantum numbers of the particle and σ is the projection of spin. In the case of electrons (or other fermions with spin $s = 1/2$) there are only two projections of spin: $s_z = 1/2$, usually denoted as α, and $s_z = -1/2$, denoted as β. The corresponding wave functions

$$\psi_{n\alpha} = \varphi_n(\mathbf{r}) \chi_\alpha; \quad \psi_{n\beta} = \varphi_n(\mathbf{r}) \chi_\beta \tag{2.20}$$

are named *spin-orbitals*.

The application of the Young operator (2.7) to a nonsymmetrized product of spin-orbitals

$$\Psi_0 = \psi_{n_1\sigma_1}(1) \psi_{n_2\sigma_2}(2) \ldots \psi_{n_N\sigma_N}(N) \tag{2.21}$$

produces the wave functions with the permutation symmetry corresponding to the Young diagram $[\lambda]$. The Pauli exclusion principle permitted only two types of the permutation symmetry corresponding to $[\lambda] = [N]$ and $[1^N]$. The wave function for the identical boson system is constructed with the aid of the Young operator $\omega^{[N]}$, which acts as the symmetrization operator $\hat{S}_N$

$$\omega^{[N]} \equiv \hat{S}_N = \frac{1}{\sqrt{N!}} \sum_P P, \tag{2.22}$$

where the summation over P runs over all the $N!$ permutations of the group π_N.

For the fermion wave functions the Young operator $\omega^{[1^N]}$ must be used, which is identical to the antisymmetrization operator $\hat{A}_N$,

$$\omega^{[1^N]} \equiv \hat{A}_N = \frac{1}{\sqrt{N!}} \sum_P (-1)^p P. \tag{2.23}$$

In Eq. (2.23) P runs over all the $N!$ permutations of the group π_N, and p is the parity of P (number of transpositions in the permutation P). The application of operator (2.23) to the nonsymmetrized product (2.21) produces the antisymmetric N-particle wave function, which is normalized, if all the one-electron functions in Eq. (2.21) are orthonormal,

$$\Psi^{[1^N]} = \frac{1}{\sqrt{N!}} \sum_P (-1)^p P \psi_{n_1\sigma_1}(1) \psi_{n_2\sigma_2}(2) \ldots \psi_{n_N\sigma_N}(N). \tag{2.24}$$

As it was first suggested by Dirac [16], for an arbitrary N-electron system, the antisymmetric wave function can be presented as a determinant. Later, Slater [17] introduced the spin functions into the determinantal representation of the electronic wave function, creating the so-called Slater's determinants. Thus,

$$\begin{aligned} \Psi^{[1^N]} &= \frac{1}{\sqrt{N!}} \sum_P (-1)^p P \psi_{n_1\sigma_1}(1) \psi_{n_2\sigma_2}(2) \ldots \psi_{n_N\sigma_N}(N) \\ &= \frac{1}{\sqrt{N!}} \begin{vmatrix} \psi_{n_1\sigma_1}(1) & \psi_{n_2\sigma_2}(1) & \ldots & \psi_{n_N\sigma_N}(1) \\ \psi_{n_1\sigma_1}(2) & \psi_{n_2\sigma_2}(2) & \ldots & \psi_{n_N\sigma_N}(2) \\ \vdots & \vdots & \ldots & \vdots \\ \psi_{n_1\sigma_1}(N) & \psi_{n_2\sigma_2}(N) & \cdots & \psi_{n_N\sigma_N}(N) \end{vmatrix}. \end{aligned} \tag{2.25}$$

The determinant equals 0, if two of its columns coincide. Thus, all electrons must have different quantum numbers, but it is just the original Pauli's formulation of his principle.

At present, the determinantal representation of N-electron function is widely used in the calculations of atoms, molecules, and solid-state structures. It is convenient for application, since one should not sum over $N!$ permutations. However, the determinantal representation has one very serious drawback. The wave function (2.25) does not correspond to a state with definite total spin S, it is only eigenfunction of the operator of its projection $\hat{S}_z$. The construction of linear combinations of determinants corresponding to the total spin S developed by Löwdin

[18] is quite cumbersome and does not allow obtaining explicit formulae for matrix elements of Hamiltonian. In next section of this chapter we discuss the method of construction of the antisymmetric wave functions that automatically correspond to the states with the definite total spin S.

2.3 The Total Wave Functions as a Product of Spatial and Spin Wave Functions

2.3.1 Two-Particle System

In previous section we assumed that each particle has its own spin function. In the case of electrons, the one-electron functions are named spin-orbitals, see Eq. (2.20). Then the total wave function was constructed in the determinantal form (2.25), which is antisymmetric, but does not describe the state with a definite value of the total spin S. However, the mentioned approach is not obligatory. The total many-electron or another many-particle wave function can be constructed as a product of a spatial wave function and a spin wave function.

The simplest case is the two-electron system. In one of the first papers devoted to just born quantum mechanics, Heisenberg [19] considered a two-electron atom and demonstrated that the total antisymmetric wave function can be constructed as a product of spatial and spin wave functions; he specially discussed the two possibilities for the total spin of the atom: $S=0$ and $S=1$. Already in the next year after Heisenberg's paper [19], Heitler and London [20] applied the same approach to the hydrogen molecule establishing the foundation for the quantum theory of a covalent bonding in molecules. It is instructive to discuss in detail the Heisenberg–Heitler–London approach in the case of the two-atom molecule.

As was stressed by Heisenberg [19], for the two-electron system there are two possibilities of constructing the antisymmetric wave functions. Namely:

$$\left.\begin{aligned}\Psi &= \Phi_{\text{sym}}(1,2)\times\Omega_{\text{antisym}}(1,2)\\ \Psi &= \Phi_{\text{antisym}}(1,2)\times\Omega_{\text{sym}}(1,2)\end{aligned}\right\}, \tag{2.26}$$

where $\Phi\,(1,2)$ is the spatial and $\Omega\,(1,2)$ is the spin two-electron wave functions.

In the theory of angular momentum in quantum mechanics it is proved that the two-electron spin wave functions for $S=0$ and $S=1$ have the following form:

$$\Omega_{\text{antisym}}^{S=0}(1,2)=\frac{1}{\sqrt{2}}\left[\chi_\alpha(1)\chi_\beta(2)-\chi_\alpha(2)\chi_\beta(1)\right] \tag{2.27}$$

and three components of the triplet spin function,

$$\Omega_{\text{sym}}^{S=1}(1,2)=\begin{cases}\frac{1}{\sqrt{2}}\left[\chi_\alpha(1)\chi_\beta(2)+\chi_\alpha(2)\chi_\beta(1)\right], & S_z=0\\ \chi_\alpha(1)\chi_\alpha(2), & S_z=1\\ \chi_\beta(1)\chi_\beta(2), & S_z=-1\end{cases}. \tag{2.28}$$

Thus, the total wave functions satisfying the Pauli principle and describing the singlet and triplet states can be constructed as

$$\begin{aligned}\Psi(S=0)&=N(S=0)[\varphi_a(1)\varphi_b(2)+\varphi_a(2)\varphi_b(1)]\Omega^{S=0}\\ \Psi(S=1)&=N(S=1)[\varphi_a(1)\varphi_b(2)-\varphi_a(2)\varphi_b(1)]\Omega^{S=1}.\end{aligned} \tag{2.29}$$

Let us find the normalization factors. They are easily found from the normalization condition

$$\begin{aligned}\langle\Psi_{S=0}|\Psi_{S=0}\rangle=1&=N_{S=0}^2\langle[\varphi_a(1)\varphi_b(2)+\varphi_a(2)\varphi_b(1)]|[\varphi_a(1)\varphi_b(2)+\varphi_a(2)\varphi_b(1)]\rangle\\ &=N_{S=0}^2\left(1+2s_{ab}^2+1\right)=N_{S=0}^2 2\left(1+s_{ab}^2\right),\end{aligned}$$

hence,

$$N_{S=0}=\frac{1}{\sqrt{2\left(1+s_{ab}^2\right)}}, \tag{2.30}$$

where the orbital overlap integral $s_{ab}=\langle\varphi_a|\varphi_b\rangle$. For the triplet state the normalization factor equals

$$N_{S=1}=\frac{1}{\sqrt{2\left(1-s_{ab}^2\right)}}. \tag{2.31}$$

In the nonrelativistic case, the Hamiltonian does not contain spin interactions; therefore, in its matrix elements the spin functions Ω disappear. The Hamiltonian for the system of two one-electron atoms, in particular for the hydrogen molecule, contains the Hamiltonians for the isolated atoms, H_a (1) and H_b (2), and the interaction operator

$$V=F+\frac{e^2}{r_{12}}+\frac{e^2}{R_{ab}}, \tag{2.32}$$

in which the last term corresponds to the nucleus–nucleus interaction and in the adiabatic approximation it is a parameter; the first term is the one-electron

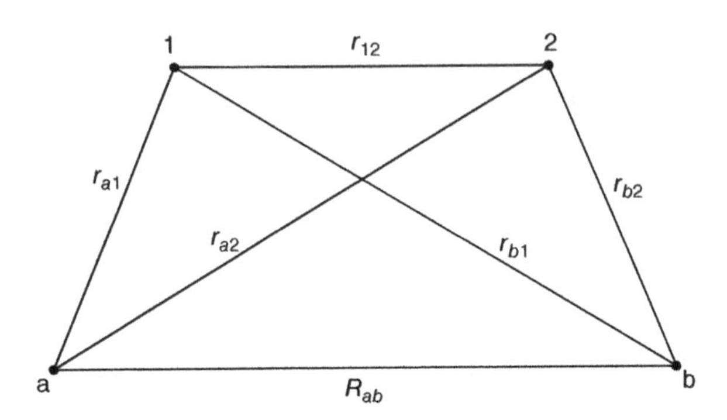

Fig. 2.2 Notations of distances in the H_2 molecule

operator describing the electron–nucleus interaction of an electron belonging to one atom with a nucleus of another atom, see notations of distances in Fig. 2.2,

$$F = f_1 + f_2 = -\frac{e^2}{r_{b1}} - \frac{e^2}{r_{a2}}. \tag{2.33}$$

The interaction energies in the singlet and triple states are expressed as the expectation values of the interaction operator that corresponds to the first order of the perturbation theory.

$$\begin{aligned} E_{\text{int}}(S=0) &= \langle \Psi(S=0)|V|\Psi(S=0)\rangle \\ &= N_{S=0}^2 \langle [\varphi_a(1)\varphi_b(2) + \varphi_a(2)\varphi_b(1)]|V|[\varphi_a(1)\varphi_b(2) + \varphi_a(2)\varphi_b(1)]\rangle, \end{aligned} \tag{2.34}$$

$$\begin{aligned} E_{\text{int}}(S=1) &= \langle \Psi(S=1)|V|\Psi(S=1)\rangle \\ &= N_{S=1}^2 \langle [\varphi_a(1)\varphi_b(2) - \varphi_a(2)\varphi_b(1)]|V|[\varphi_a(1)\varphi_b(2) - \varphi_a(2)\varphi_b(1)]\rangle. \end{aligned} \tag{2.35}$$

The results depend only upon spatial wave functions, but they are different for different total spin due to the different permutation symmetry of the two-electron spatial wave functions: symmetric for the singlet state and antisymmetric for the triplet state.

The expression for the interaction energy in the singlet state (2.34) can be transposed as

$$E_{\text{int}}(S=0) = N_{S=0}^2\{\langle\varphi_a(1)\varphi_b(2)|V|\varphi_a(1)\varphi_b(2)\rangle + \langle\varphi_a(1)\varphi_b(2)|V|\varphi_a(2)\varphi_b(1)\rangle$$
$$+ \langle\varphi_a(2)\varphi_b(1)|V|\varphi_a(1)\varphi_b(2)\rangle + \langle\varphi_a(2)\varphi_b(1)|V|\varphi_a(2)\varphi_b(1)\rangle\}$$
$$= \frac{1}{\left(1+s_{ab}^2\right)}(V_{ab,ab} + V_{ab,ba}), \tag{2.36}$$

where we denoted

$$\begin{aligned} V_{ab,ab} &= \langle\varphi_a(1)\varphi_b(2)|V|\varphi_a(1)\varphi_b(2)\rangle, \\ V_{ab,ba} &= \langle\varphi_a(1)\varphi_b(2)|V|\varphi_b(1)\varphi_a(2)\rangle. \end{aligned} \tag{2.37}$$

Using the expression (2.32) for V, the matrix element $V_{ab,ab}$ is presented as two terms

$$V_{ab,ab} = \langle\varphi_a(1)\varphi_b(2)|F|\varphi_a(1)\varphi_b(2)\rangle + \left\langle \varphi_a(1)\varphi_b(2)\left|\frac{e^2}{r_{12}}\right|\varphi_a(1)\varphi_b(2)\right\rangle, \tag{2.38}$$

where the two-electron matrix element

$$\alpha_{ab} = \left\langle \varphi_a(1)\varphi_b(2)\left|\frac{e^2}{r_{12}}\right|\varphi_a(1)\varphi_b(2)\right\rangle = \int \varphi_a(1)^*\varphi_a(1)\frac{e^2}{r_{12}}\varphi_b(2)^*\varphi_b(2)dV_1dV_2 \tag{2.39}$$

is named the *Coulomb integral*, we denote it as α_{ab}. Coulomb integral is the energy of the Coulomb interaction of two electron densities distributed in space, localized each on its atoms: $e|\varphi_a(1)|^2$ and $e|\varphi_b(2)|^2$. It has the same sense, as the Coulomb energy in classical physics, but instead of point charges it contains distributed charges.

The matrix element $V_{ab,ba}$ is presented similar to $V_{ab,ab}$ as

$$V_{ab,ba} = \langle\varphi_a(1)\varphi_b(2)|F|\varphi_b(1)\varphi_a(2)\rangle + \left\langle \varphi_a(1)\varphi_b(2)\left|\frac{e^2}{r_{12}}\right|\varphi_b(1)\varphi_a(2)\right\rangle, \tag{2.40}$$

where the two-electron matrix element

$$\beta_{ab} = \left\langle \varphi_a(1)\varphi_b(2)\left|\frac{e^2}{r_{12}}\right|\varphi_b(1)\varphi_a(2)\right\rangle = \int \varphi_a(1)^*\varphi_b(1)\frac{e^2}{r_{12}}\varphi_b(2)^*\varphi_a(2)dV_1dV_2 \tag{2.41}$$

is named the *exchange integral*, we denote it as β_{ab}. Exchange integral is the energy of the Coulomb interaction of two electron densities delocalized between atoms: $e\varphi_a(1)^*\varphi_b(1)$ and $e\varphi_b(2)^*\varphi_a(2)$. The origin of the exchange energy is in the antisymmetry of the total wave function, that is, in the Pauli exclusion principle. It is a specific quantum mechanical effect. The exchange energy is equal to zero, if the overlap of orbitals φ_a and φ_b is negligible.

So the interaction energy (2.36) is expressed as

$$E_{\text{int}}(S=0)=\frac{1}{1+s_{ab}^2}[F_{ab,ab}+F_{ab,ba}+\alpha_{ab}+\beta_{ab}]. \tag{2.42}$$

Let us write the explicit expressions for one-electron integrals, taking into account the expression (2.33) for F and the identity of two hydrogen atoms,

$$\begin{aligned}F_{ab,ab}&=\langle\varphi_a(1)\varphi_b(2)|f_1+f_2|\varphi_a(1)\varphi_b(2)\rangle\\&=\langle\varphi_a(1)|f_1|\varphi_a(1)\rangle+\langle\varphi_b(2)|f_2|\varphi_b(2)\rangle=2\langle\varphi_a|f|\varphi_a\rangle=2f_{aa}\end{aligned} \tag{2.43}$$

$$\begin{aligned}F_{ab,ba}&=\langle\varphi_a(1)\varphi_b(2)(f_1+f_2)\varphi_b(1)\varphi_a(2)\rangle=\\&=\langle\varphi_a(1)|f_1|\varphi_b(1)\rangle\langle\varphi_b(2)|\varphi_a(2)\rangle+\langle\varphi_a(1)|\varphi_b(1)\rangle\langle\varphi_b(2)|f_2|\varphi_a(2)\rangle\\&=2f_{ab}s_{ab}\end{aligned} \tag{2.44}$$

and

$$F_{ab,ab}+F_{ab,ba}=2(f_{aa}+s_{ab}f_{ab}). \tag{2.45}$$

The final expression for the interaction energy in the singlet state is

$$E_{\text{int}}(S=0)=\frac{1}{1+s_{ab}^2}[2(f_{aa}+s_{ab}f_{ab})+\alpha_{ab}+\beta_{ab}]. \tag{2.46}$$

The interaction energy in the triplet state is obtained in the similar manner,

$$E_{\text{int}}(S=1)=\frac{1}{1-s_{ab}^2}[2(f_{aa}-s_{ab}f_{ab})+\alpha_{ab}-\beta_{ab}]. \tag{2.47}$$

Thus, we see that even in the absence of the spin interactions, the energy depends on the value of the total spin. This is due to the Pauli exclusion principle. In the case of the antisymmetric total wave function, the symmetry of the spatial wave function depends on the value of the total spin. As a result, exchange and overlap integrals appear in the interaction energy for the states with $S=0$ and $S=1$ with different signs. This produces two qualitatively different potential curves, see

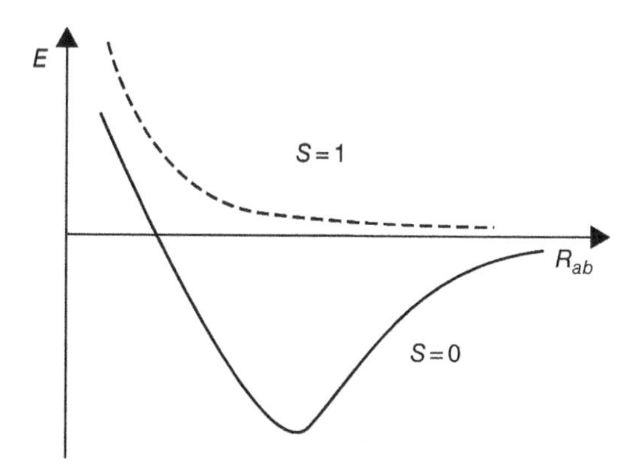

Fig. 2.3 Potential energy curves for singlet and triplet states of the hydrogen molecule

Fig. 2.3. In the singlet state the molecule is stable because of the redistributed electron density in the space between atoms (*covalent bonding*). In the triplet state the potential curve is repulsive. Thus, the coupled spins produce the stable molecule and the parallel spins destabilize the chemical bonding.

However, it is important to stress once more that these effects are not connected with the spin interactions, they stem from the different permutation symmetry of the spatial functions in the singlet and triplet states and have a pure electrostatic origin. At large interatomic distance the overlap of atomic orbitals becomes negligible and $E_{\text{int}}(S=0) = E_{\text{int}}(S=1)$, the interaction energy does not depend on the value of the total spin.

2.3.2 General Case of N-Particle System

Though the concept of spin has enabled to explain the nature of chemical bond, electron spins are not involved directly in the formation of the latter. The interaction responsible for chemical bonding has a purely electrostatic nature. If the spin interaction can be neglected in the molecular Hamiltonian, the latter commutes with the operator of the total spin squared, $\hat{S}^2$. Consequently, the total spin S is a good quantum number, and the wave functions should be eigenfunctions of $\hat{S}^2$. As for the two-particle system, in general the N-particle case, the spatial coordinates and spin variables in the electron wave function may be separated, and the latter can be presented as a product of a spatial wave function and a spin wave function. As we will show below, there is a unique correspondence between the permutation symmetry of the spatial N-particle wave function and the value of the total spin S.

In the system of N particles each possessing the spin s, the coupling of the particle spins is performed according to general rules of the vector addition of angular momenta in quantum mechanics. The coupling of spins $s = 1/2$ is the most simple case. For two electrons (let us consider, for definability, electrons) there are only two possibilities, as we already discussed in Section 2.3.1:

$$S = \begin{cases} \uparrow\ \downarrow = 0 \\ \uparrow\ \uparrow = 1 \end{cases} \tag{2.48}$$

Two particles with $s = 1/2$ can be in the singlet state or in the triplet state. The latter, in correspondence with its name, is triply degenerate, $2S + 1 = 3$, with three values of the spin projection $S_z = \pm 1, 0$.

Three particles can also be only in two states: the doublet state with $S = 1/2$ and the quartet state with $S = 3/2$, although the doublet state can be constructed by two ways:

$$S = \begin{cases} \uparrow\ \downarrow\ +\ \uparrow = \frac{1}{2} \\ \uparrow\ \uparrow\ +\ \downarrow = \frac{1}{2} \\ \uparrow\ \uparrow\ +\ \uparrow = \frac{3}{2} \end{cases} \tag{2.49}$$

Thus, except for the trivial degeneracy $(2S + 1)$ of the state with the total spin S, there is a spin degeneracy equal to the number of ways of coupling the spins of individual particles that produces the definite value of the total spin. These numbers can be conveniently found from the so-called spin coupling diagram, see Fig. 2.4.

The total wave function of an arbitrary electron system must satisfy three symmetry requirements. It should be:

1. antisymmetric with respect to electron permutations (obey the Pauli exclusion principle);
2. an eigenfunction of $\hat{S}^2$, which means that it should belong to one of the irreducible representations of the spin-space rotation group;
3. a basic function of one of the irreducible representations $\Gamma^{(\alpha)}$ of the point symmetry group of the molecule.

The third condition is satisfied by using the projection operator for a point group similar to the Young operators (2.7), see Appendix A. In order to satisfy the first two conditions, instead of following the cumbersome procedure of constructing the corresponding linear combinations of determinants, it is convenient to use the permutation group technique. By presenting the total wave function as a product of a spatial and a spin wave function, symmetrized according to the irreducible

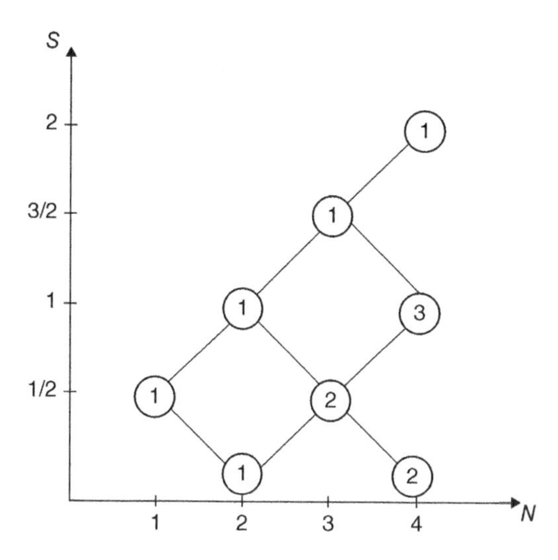

Fig. 2.4 The spin diagram for $s = 1/2$; numbers in circles give the spin degeneracy for a given S and N

representations $\Gamma^{[\lambda]}$ of the permutation group, one is able to satisfy the first two requirements automatically. The Young diagrams of the spatial and spin wave functions should obey a condition that ensures that the total wave function is antisymmetric. As it is proved in Appendix B, such an antisymmetric function is

$$\Psi_t^{[1^N]} = \frac{1}{\sqrt{f_\lambda}} \sum_r \Phi_{rt}^{[\lambda]} \Omega_{\tilde{r}}^{[\tilde{\lambda}]}, \tag{2.50}$$

where $\Gamma^{[\tilde{\lambda}]}$ denotes the representation *conjugate* (it is named also *adjoint*) to $\Gamma^{[\lambda]}$. Its matrix elements are $\Gamma_{\tilde{r}\tilde{t}}^{[\tilde{\lambda}]}(P) = (-1)^p \Gamma_{rt}^{[\lambda]}(P)$. The Young diagram $[\tilde{\lambda}]$ is dual to $[\lambda]$, that is, it is obtained from the latter by replacing rows by columns. For example,

The sum in (2.50) is taken over all the basis functions of the representation, the index t enumerates the different bases of the representation $\Gamma^{[\lambda]}$, and the factor $1/\sqrt{f_\lambda}$ ensures the correct normalization of $\Psi_t^{[1^N]}$.

It is easy to ascertain that the function (2.50) is antisymmetric. Let us act upon (2.50) with an arbitrary permutation $P \in \pi_N$. By using the properties of the basis functions of an irreducible representation, we obtain

$$P\Psi_t^{[1^N]} = \frac{1}{\sqrt{f_\lambda}} \sum_r \sum_u \sum_v \Gamma_{ur}^{[\lambda]}(P) \Gamma_{\tilde{v}\tilde{r}}^{[\tilde{\lambda}]}(P)\, \Phi_{ut}^{[\lambda]} \Omega_{\tilde{v}}^{[\tilde{\lambda}]}.$$

By taking into account the definition of the conjugate representation $\Gamma_{\tilde{v}\tilde{r}}^{[\tilde{\lambda}]}(P) = (-1)^p \Gamma_{vr}^{[\lambda]}$ and the orthogonality properties of the representation matrix, we find

$$\begin{aligned} P\Psi_t^{[\lambda]} &= \frac{1}{\sqrt{f_\lambda}} \sum_{uv} \left(\sum_r \Gamma_{ur}^{[\lambda]}(P) \Gamma_{vr}^{[\lambda]}(P) \right) (-1)^p \Phi_{ut}^{[\lambda]} \Omega_{\tilde{v}}^{[\tilde{\lambda}]} \\ &= \frac{1}{\sqrt{f_\lambda}} \sum_u (-1)^p \Phi_{ut}^{[\lambda]} \Omega_{\tilde{u}}^{[\tilde{\lambda}]} = (-1)^p \Psi_t^{[\lambda]}, \end{aligned} \tag{2.51}$$

which demonstrates the correctness of Eq. (2.50).

The construction of the spatial functions $\Phi_{rt}^{[\lambda]}$ is described in Section 2.2. Acting by the Young operator (2.7) on the antisymmetrized product of orthonormal set of one-particle spatial functions (2.8) we obtain the normalized function (2.9), which is transformed in accordance with the representation $\Gamma^{[\lambda]}$.

The spin function, $\Omega \tilde{r}^{[\tilde{\lambda}]}$, can be found in a similar manner by applying the operator $\omega_{\tilde{r}\tilde{t}}^{[\tilde{\lambda}]}$ to a product of one-electron spin functions χ_σ. χ_σ is not a continuous function, but a two-component quantity. These two components, $\chi_\alpha(\sigma = 1/2)$ and $\chi_\beta(\sigma = -1/2)$, make up a *spinor* (see section 56 in book by Landau and Lifshitz [21]). With rotation of the coordinate system, the components of spinors transform in such a way that the bilinear combination $\chi_\alpha \eta_\beta - \chi_\beta \eta_\alpha$ remains invariant for any two spinors χ and η. A spinor may be regarded as a vector in a two-dimensional space with the metric given by the antisymmetric metric tensor

$$g^{\mu\nu} = \begin{pmatrix} 0 & 1 \\ -1 & 0 \end{pmatrix}. \tag{2.52}$$

Then the scalar product of two spinors can be found from the general formulas of tensor calculus:

$$(\chi, \eta) = \sum_{\mu\nu} g^{\mu\nu} \chi_\mu \eta_\nu = \chi_1 \eta_2 - \chi_2 \eta_1. \tag{2.53}$$

The set of 2^N quantities

$$\Omega_{\sigma_1\sigma_2\ldots\sigma_N} = \chi_{\sigma_1}(1)\chi_{\sigma_2}(2)\ldots\chi_{\sigma_N}(N) \tag{2.54}$$

forms a tensor of rank N in the two-dimensional spinor space. Rotations of the coordinate system transform the components (2.54) according to a 2^N-dimensional representation of the rotation group. The latter representation can be decomposed into irreducible parts by constructing symmetrized linear combinations of the above components making use of the Young operators (see Appendix C, Section C.3.2). Each component in Eq. (2.54) is characterized by certain values of the total electron spin projection, $M_S = \sum_{i=1}^{N} \sigma_i$. The action of the Young operator $\omega_{\tilde{r}\tilde{t}}^{[\tilde{\lambda}]}$ upon this tensor of rank N in the two-dimensional spinor space, Eq. (2.54), transforms it into the functions:

$$\Omega_{\tilde{r},\sigma_1\ldots\sigma_N}^{[\tilde{\lambda}]} = \omega_{\tilde{r}\tilde{t}}^{[\tilde{\lambda}]} \Omega_{\sigma_1\ldots\sigma_N}. \tag{2.55}$$

For fixed $\sigma_1\sigma_2 \ldots \sigma_N$, these functions transform according to a certain representation $\Gamma^{[\tilde{\lambda}]}$ of the permutation group of spin variables. For a fixed $\tilde{r}$ they transform according to a certain representation of the special unitary group in a two-dimensional space, $\mathbf{SU}_2$, see Section C.1.2. The latter group is isomorphic to the rotation group in the three-dimensional space $\mathbf{R}_3$ and, consequently, there is a one-to-one correspondence between the irreducible representations of $\mathbf{SU}_2$ and the irreducible representations $D^{(S)}$ of the three-dimensional rotation group, that is, with the value of spin S. To obtain an independent basic set for the representation $D^{(S)}$, it is sufficient to select any fixed $\tilde{t}$ at which the action of $\omega_{\tilde{r}\tilde{t}}^{[\tilde{\lambda}]}$ on $\Omega_{\sigma_1\sigma_2\ldots\sigma_N}$ does not give zero (this is why we have omitted the index $\tilde{t}$ in $\Omega_{\tilde{r},\sigma_1\ldots\sigma_N}^{[\tilde{\lambda}]}$ in Eq. (2.55)).

The connection between the Young diagram $[\tilde{\lambda}]$ and the value of S can be very easily found from the following simple considerations. Since any antisymmetric function is zero whenever two of its arguments are the same, the spin diagrams for particles with spin 1/2 cannot have more than two boxes in a column, that is, each diagram has no more than two rows. Consequently, the coordinate Young diagrams $[\lambda]$ dual to it cannot have more than two columns. If in one box of a column in a spin Young diagram, the electron spin projection is 1/2, then in the other box of this column the electron spin projection is $-1/2$, that is, the spins of these two electrons should be coupled. It is evident that the contribution to the total spin of the system will come only from uncoupled electron spins; their number equals the difference between the lengths of the rows in the corresponding Young diagram, $\left(\tilde{\lambda}^{(1)} - \tilde{\lambda}^{(2)}\right)$, and we arrive at the simple formula

$$S = \frac{1}{2}\left(\tilde{\lambda}^{(1)} - \tilde{\lambda}^{(2)}\right). \tag{2.56}$$

Equation (2.56) enables one to find the values of the spin, S, for each spin Young diagram. For example, for the Young diagram $[\tilde{\lambda}] = [31]$ the spin is $S = 1$,[2]

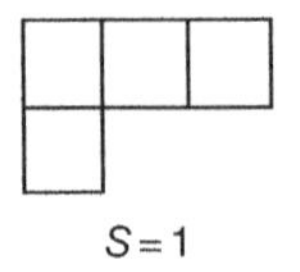

$S = 1$

Let us consider, as an example, the construction of spin functions for a system of three electrons in a state with the total spin $S = 1/2$ (i.e., for the representation $\Gamma^{[\tilde{2}\tilde{1}]}$). The matrices $\Gamma^{[\tilde{2}\tilde{1}]}(P)$ differ from the matrices $\Gamma^{[21]}(P)$, Eq. (2.16), by the factor $(-1)^p$. For the function corresponding to the total spin projection $M_S = 1/2$, $\Omega_{\sigma_1\sigma_2\sigma_3}$ in Eq. (2.54) equals

$$\Omega_{\frac{1}{2}\frac{1}{2}-\frac{1}{2}} = \alpha(1)\alpha(2)\beta(3), \tag{2.54a}$$

where to keep our notations simple, we have denoted the spin functions with spin projection $+1/2$ and $-1/2$ as α and β, respectively. The result of the action of the operator $\omega_{\tilde{r}\tilde{1}}^{[21]}$ on the product $\alpha(1)\alpha(2)\beta(3)$ is zero, so as a projection operator we should take:

$$\omega_{\tilde{r}\tilde{2}}^{[21]} = \sqrt{\frac{2}{3!}}\sum_P \Gamma_{\tilde{r}\tilde{2}}^{[21]}(P)P. \tag{2.57}$$

The result of the action by the operator (2.57) on the function (2.54a) is

$$\begin{aligned} \Omega_{\tilde{1},\alpha\alpha\beta}^{[21]} &= [-\alpha(1)\alpha(3)\beta(2) + \alpha(2)\alpha(3)\beta(1)], \\ \Omega_{\tilde{2},\alpha\alpha\beta}^{[21]} &= \frac{1}{\sqrt{3}}[2\alpha(1)\alpha(2)\beta(3) - \alpha(1)\alpha(3)\beta(2) - \alpha(2)\alpha(3)\beta(1)]. \end{aligned} \tag{2.58}$$

The obtained spin functions are nonnormalized. For their normalization we must multiply functions (2.58) on the factor $1/\sqrt{2}$,

$$\begin{aligned} \Omega_{\tilde{1},\alpha\alpha\beta}^{[21]} &= \frac{1}{\sqrt{2}}[-\alpha(1)\beta(2)\alpha(3) + \beta(1)\alpha(2)\alpha(3)], \\ \Omega_{\tilde{2},\alpha\alpha\beta}^{[21]} &= \frac{1}{\sqrt{6}}[2\alpha(1)\alpha(2)\beta(3) - \alpha(1)\beta(2)\alpha(3) - \beta(1)\alpha(2)\alpha(3)]. \end{aligned} \tag{2.58a}$$

[2] Such the one-to-one correspondence between a Young diagram and the total spin exists only for particles with spin 1/2. For particles with $s > 1/2$ several values of S may correspond to the permutation symmetry of given Young diagram. For instance, if the spin of particle equal to 1 then in case of the Young diagram [31], the corresponding values of S are 1, 2 and 3, see Appendix C, Table 2 in Section C4.

This illustrated the problem of normalization in the case of spin functions. Both functions in Eq. (2.58a) enter into one linear combination to make up the total wave function. They therefore describe the same state of the system of three electrons with the total spin $S = 1/2$ and its projection $M_S = 1/2$.

Let us write the general nonsymmetrized spin function (2.54) in a form:

$$\Omega_{N,S_z} = \alpha(1)\alpha(2)\ldots\alpha(N_\alpha)\beta(N_\alpha + 1)\ldots\beta(N_\alpha + N_\beta) \tag{2.59}$$

The function (2.59) corresponds to $S_z = (1/2)(N_\alpha - N_\beta)$, on the other hand, $N = N_\alpha + N_\beta$. From these two equations it follows that

$$N_\alpha = \frac{N}{2} + S_z, \quad N_\beta = \frac{N}{2} - S_z. \tag{2.60}$$

and function (2.59) really depends only on N and S_z. The action by $N!$ permutations on function (2.59) results in

$$\frac{N!}{N_\alpha! N_\beta!} = \frac{N!}{(N/2 + S_z)!(N/2 - S_z)!} \tag{2.61}$$

different (not similar) terms; each term repeating $(N/2 + S_z)!(N/2 - S_z)!$ times. In the case of the maximum spin $S_{\max} = N/2$, which corresponds to the one-dimensional representation $[N]$, all matrix elements equal to 1 and the normalization factor is given by the number of orthogonal terms, Eq. (2.61), if we sum over the permutations between α and β spin projections, $P_{\alpha\beta}$. Thus,

$$\Omega_{S_{\max},S_z} = \sqrt{\frac{(N/2 + S_z)!(N/2 - S_z)!}{N!}} \sum_{P_{\alpha\beta}} P_{\alpha\beta} \Omega_{N,S_z} \tag{2.62}$$

In a general case of multidimensional representations, the spin normalization factor was recently obtained by Yurovsky [22]. According to his derivation,

$$\Omega_{r,S,S_z} = C_{S,S_z} \sum_P \Gamma^{[\lambda]}_{rt_0}(P) P \Omega_{N,S_z}, \tag{2.63}$$

where the summation over P runs over all $N!$ permutations of the group π_N, t_0 is the *fundamental* Young tableau, that is, the tableau in which the numbers are in their natural order reading along the rows and from top to bottom, and the normalization factor

$$C_{S,S_z} = \frac{1}{(N/2 + S_z)!(N/2 - S)!} \sqrt{\frac{(2S + 1)(S + S_z)!}{(N/2 + S + 1)(2S)!(S - S_z)!}}. \tag{2.64}$$

In the case of the maximum spin $S = N/2$, Eqs. (2.63) and (2.64) are transformed into Eq. (2.62), if we reduce the sum over P to the sum over $P_{\alpha\beta}$.

Thus, presenting the total wave function as a product (2.50) of a spatial and a spin wave function, symmetrized according to the appropriate irreducible representations of the permutation group, enables one, on the one hand, to construct the wave function as an eigenfunction of $\hat{S}^2$ and $\hat{S}_z$; on the other hand, to satisfy the antisymmetry requirement automatically. At the same time this approach is very convenient in practical calculations. For instance, it allows obtaining closed formulae for the matrix elements of spin-independent Hamiltonian in state with a given value of total spin, see Refs. [13, 23]. In the problems that do not involve spin interactions, the calculations made with the spatial wave functions $\Phi_{rt}^{[\lambda]}$ are much more natural and physical than those made with functions containing spin. The form of the spin wave functions $\Omega_{\tilde{r}}^{[\tilde{\lambda}]}$ is irrelevant, since they do not enter into the final result of calculations of the matrix elements of spin-independent operators. Nevertheless, the results depend on the value of the total spin, since the "memory" of the spin is retained in the permutation symmetry $[\lambda]$ of the spatial wave function. This approach has been named *quantum chemistry without spin* and was independently developed by the author [13, 24, 25] and Matsen [26–28]. See discussion in Chapter 1, p. 13, and a comprehensive review by Klein [29].

As we discussed above, in nonrelativistic problems the spatial and spin coordinates can be separated, and the quantum states may be classified according to their total electron spin S and the corresponding irreducible representation $\Gamma^{(\alpha)}$ of the point symmetry group of the system. Such states are called the *molecular multiplets* and are denoted as ${}^{2S+1}\Gamma^{(\alpha)}$. The energy level of a multiplet is $f_\alpha(2S+1)$ times degenerated, where f_α is the dimension of the representation $\Gamma^{(\alpha)}$. However, not all the $(S, \Gamma^{(\alpha)})$ pairs are allowed by the Pauli principle, since the corresponding total wave functions may not be antisymmetric. An elegant approach of finding the molecular multiplets based on the representation of the total wave function in the form (2.50) was developed in Refs. [13, 30, 31] and discussed in Chapter 4.

It should be mentioned that the application of the total electron wave function in the form (2.50), namely,

$$\Psi_t^{[1^N]}(S, S_z) = \frac{1}{\sqrt{f_\lambda}} \sum_r \Phi_{rt}^{[\lambda]} \Omega_{\tilde{r}, S, S_z}^{[\tilde{\lambda}]} \tag{2.65}$$

is also useful in those cases where one deals with characteristics of a system related to spin. Usually the spin interactions can be considered as small in comparison with electrostatic ones and can be treated by the perturbation theory. The Hamiltonian of the system is presented as

$$H = H_0 + V_S, \tag{2.66}$$

where H_0 is a spin-free Hamiltonian and V_S is a small perturbation depending on spin. It can be spin–orbit or spin–spin interactions. These interactions can be found in the first order of the perturbation theory as the expectation value of the operator V_S in the unperturbed state, which is described by functions (2.65),

$$E_S^{(1)} = \left\langle \Psi_t^{[1^N]}(S,S_z) \middle| V_S \middle| \Psi_t^{[1^N]}(S,S_z) \right\rangle. \tag{2.67}$$

Similar methods can be used for the description of quantum many-body states of spinor atomic gases, see Ref. [32].

References

[1] V.L. Lyuboshitz and M.I. Podgoretskii, *Sov. Phys. JETP* **28**, 469 (1969).
[2] V.L. Lyuboshitz and M.I. Podgoretskii, *Sov. Phys. JETP* **30**, 91, 100 (1970).
[3] I.G. Kaplan and A.P. Markin, *Sov. Phys. Dokl.* **14**, 36 (1969).
[4] D. Akoury et al., *Science* **318**, 949 (2007).
[5] U. Miranda and I.G. Kaplan, *Eur. Phys. J. D* **63**, 263 (2011).
[6] E. Wigner and E.E. Witmer, *Z. Phys.* **51**, 859 (1928).
[7] R.S. Mulliken, *Rev. Mod. Phys.* **4**, 1 (1932).
[8] G. Herzberg, *Molecular Spectra and Molecular Structure, I. Spectra of Diatomic Molecules*, D. Van Nostrand, Princeton, New Jersey, 1950.
[9] J.B. Hartle and J.R. Taylor, *Phys. Rev.* **178**, 2043 (1969).
[10] A. Casher, G. Frieder, M. Glűck, and A. Peres, *Nucl. Phys.* **66**, 632 (1965).
[11] R. Mirman, *Nuovo Cimento* **18B**, 110 (1973).
[12] M. Hamermesh, *Group Theory*, Addison Wesley, Reading, Massachusetts, 1964.
[13] I.G. Kaplan, *Symmetry of Many-Electron Systems*, Academic Press, New York, 1975 (translation by J. Gerratt of enlarged Russian Edition, Nauka, Moscow, 1969).
[14] D.E. Rutherford, *Substitutional Analysis*, Edinburgh University Press, Edinburgh, 1947.
[15] F.D. Murnaghan, *The Theory of Group Representations*, John Hopkins Press, Baltimore, Maryland, 1938.
[16] P.A.M. Dirac, *Proc. Roy. Soc. Lond. A* **112**, 621 (1926).
[17] J.C. Slater, *Phys. Rev.* **34**, 1293 (1929).
[18] P.O. Löwdin, *Phys. Rev.* **97** 1509 (1955).
[19] W. Heisenberg, *Z. Phys.* **39**, 499 (1926).
[20] W. Heitler and F. London, *Z. Phys.* **44**, 455 (1927).
[21] L.D. Landau and E.M. Lifshitz, *Quantum Mechanics (Nonrelativistic Theory)*, 3rd edn., Pergamon Press, Oxford, 1977.
[22] V.A. Yurovsky, *Int. J. Quantum Chem.* **113**, 1436 (2013).
[23] I.G. Kaplan and O.B. Rodimova, *Int. J. Quantum Chem.* **7**, 1203 (1973).
[24] I.G. Kaplan, *Sov. Phys. JETP* **17**, 261 (1963).
[25] I.G. Kaplan, *Theor. Exp. Chem.* **1**, 399, 407 (1965); 2, 335 (1966); 3, 83 (1967).
[26] F.A. Matsen, *Adv. Quantum Chem.* **1**, 60 (1964).
[27] F.A. Matsen, *J. Phys. Chem.* **68**, 3282 (1964).
[28] F.A. Matsen, *J. Am. Chem. Soc.* **92**, 3525 (1970).
[29] D.J. Klein, *Int. J. Quantum Chem.* **111**, 76 (2011).
[30] I.G. Kaplan, *Sov. Phys. JETP* **10**, 747 (1960).
[31] I.G. Kaplan and O.B. Rodimova, *Sov. Phys. JETP* **39**, 764 (1974).
[32] V.A. Yurovsky, *Phys. Rev. Lett.* **113**, 200406 (2014).

3

Can the Pauli Exclusion Principle Be Proved?

3.1 Critical Analysis of the Existing Proofs of the Pauli Exclusion Principle

The general formulation of the Pauli exclusion principle represented in Section 1.1 can be considered from two viewpoints. On the one hand, it asserts that particles with half-integer spin (fermions) are described by antisymmetric wave functions and particles with integer spin (bosons) are described by symmetric wave functions. This is the so-called spin–statistics connection. The reasons why this connection between the value of spin and the permutation symmetry of wave function exists are still unknown. In Section 1.2, p. 19, we discuss the existing unsatisfactory situation with numerous "proofs" of the spin–statistics connection. There, I cited references devoted to this problem, including the comprehensive book by Duck and Sudarshan [1] and their review [2] where they criticized all proofs of the spin–statistics connection published at that time. They assumed as correct only an old Sudarshan proof [3]. However, Wightman [4] in his review on the Duck–Sudarshan book [1] noted that the proof [3] is also incorrect. It is interesting to mention that Robert Feynman in his famous lectures on physics apologized in front of audience that he cannot give an elementary explanations why the spin–statistics connection exists, see Ref. [5]. He noted:

> It appears to be one of the few places in physics where there is a rule which can be stated very simply, but for which no one found a simple and easy explanation.

The Pauli Exclusion Principle: Origin, Verifications, and Applications, First Edition. Ilya G. Kaplan.

We should state that to the best of our knowledge, there still has been no progress in this direction; see also footnote 9 in Chapter 1. In what follows we will not discuss more this very important but unsolved problem.

Let us turn to another aspect of the Pauli exclusion principle. According to it, only two types of permutation symmetry are allowed: symmetric and antisymmetric. Both belong to the one-dimensional representations of the permutation group, while all other types of permutation symmetry are forbidden. On the other hand, the Schrödinger equation is invariant under any permutation of identical particles and its solutions may belong to any representation of the permutation group, including the multidimensional representations that are forbidden by the Pauli exclusion principle. Here we can repeat the question asked in the end of Section 1.2: *whether the Pauli principle limitation on the solutions of the Schrödinger equation follows from the fundamental principles of quantum mechanics or it is an independent principle*.

Depending on the answer to this question, physicists can be divided into two groups.

Some physicists, including the founders of quantum mechanics Pauli [6] and Dirac [7] (see also books by Schiff [8] and Messiah [9]), had assumed that there are no laws in Nature that forbid the existence of particles described by wave functions with more complicated permutation symmetry than those of bosons and fermions, and that the existing limitations are only due to the specific properties of the known elementary particles.

Messiah [9, 10] has even introduced the term *symmetrization postulate* to emphasize the primary nature of the constraint on the allowed types of the wave function permutation symmetry. However, before Messiah, the independence of his exclusion principle from other fundamental principles of quantum mechanics was stressed by Pauli in his Princeton address [6]:

> The exclusion principle could not be deduced from the new quantum mechanics but remains an independent principle which excludes a class of mathematically possible solutions of the wave equation

In fact, the existence of permutation degeneracy should not introduce additional uncertainty into characteristic of the state. From the Wigner–Eckart theorem formulated for the permutation group, see Eq. (D.16) in Appendix D, it follows that the matrix element of an operator $\hat{L}$, which is symmetric in all the particles, can be presented as

$$\left\langle \Psi_r^{[\lambda]} \middle| \hat{L} \middle| \Psi_{\bar{r}}^{[\lambda]} \right\rangle = \delta_{r\bar{r}} \left\langle \Gamma^{[\lambda]} \middle\| \hat{L} \middle\| \Gamma^{[\lambda]} \right\rangle, \tag{3.1}$$

where the index r labels the basis functions of the representation $\Gamma^{[\lambda]}$ of the permutation group and $[\lambda]$ is the Young diagram (see Appendix B). The double vertical line on the right-hand side of the formula means that the matrix element is independent on the basis function index. Thus, the expectation value of operator L is the same for all functions belonging to the degenerate state described by an arbitrary irreducible representation $\Gamma^{[\lambda]}$ of the permutation group (nevertheless, in the degenerate states some contradictions with the concept of particle identity are appeared, see Section 3.2).

There is another viewpoint on this problem; according to it, the symmetry postulate is not an independent principle and can be derived from fundamental principles of quantum mechanics, in particular, from the principle of indistinguishability of identical particles. This idea is represented not only in articles (see critical comments in Refs. [10, 11]) but also in textbooks [12–14], including the famous textbook by Landau and Lifshitz [13] translated into many languages. The incorrectness of the proof by Corson [12] was mentioned by Girardeau [11], and the proofs represented in Refs. [12–14] were critically analyzed in my first paper on the Pauli exclusion principle [15] (a more detailed criticism was given in Refs. [16–18]); nevertheless, incorrect proofs of the Pauli principle still appear in current literature; see, for instance, the review [19].[1] Even in the recently published book by Piela [21], the proof of the Pauli exclusion principle is represented with the same errors as in the textbooks cited above. Thus, it is worthwhile to discuss this matter once more in this book.

The typical argumentation (it is similar in Refs. [12–14, 19, 21]) is the following. From the requirement that the states of a system obtained by permutations of identical particles must all be physically equivalent, one concludes that the transposition of any two identical particles should multiply the wave function only on an insignificant phase factor,

$$P_{12}\Psi(x_1,x_2) = \Psi(x_2,x_1) = e^{i\alpha}\Psi(x_1,x_2), \tag{3.2}$$

where α is a real constant and x is the set of spatial and spin variables. One more application of the permutation operator P_{12} gives

$$\Psi(x_1,x_2) = e^{i2a}\Psi(x_1,x_2) \tag{3.3}$$

or

$$e^{2i\alpha} = 1 \quad \text{and} \quad e^{i\alpha} = \pm 1. \tag{3.4}$$

Since all particles are assumed to be identical, the wave function should change in the exactly same way under transposition of any pair of particles, that is, it should be either totally symmetric or totally antisymmetric.

[1] The "proof" in Ref. [19] followed from the conception accepted in many early studies in the fractional statistics area that the Hilbert space in the case of identical particles must be one-dimensional. This conception was based on the wrong statement in the paper by Mirman [20]; for a more detailed discussion, see Section 5.4.1 devoted to the fractional statistics.

This proof contains two evident incorrectnesses at once. The first one simply follows from the group theory. Namely Eq. (3.2) is valid only for the one-dimensional representations. The application of a group operation to one of basis functions, belonging to some multidimensional representation, transforms it in a linear combination of basis functions of this representation. Namely,

$$P_{12}\Psi_i = \sum_k \Gamma_{ki}(P_{12})\Psi_k. \tag{3.5}$$

The application of the permutation operator P_{12} to both sides of Eq. (3.5) leads to the identity:

$$\begin{aligned} P_{12}\{P_{12}\Psi_i\} = \Psi_i = P_{12}\sum_k \Gamma_{ki}(P_{12})\Psi_k = \sum_l \left[\sum_k \Gamma_{lk}(P_{12})\Gamma_{ki}(P_{12})\right]\Psi_l \\ = \sum_l \Gamma_{li}(P_{12}^2 = I)\Psi_l = \Psi_i. \end{aligned} \tag{3.6}$$

Using this identity, we cannot arrive at any information about the symmetry in contrary with Eq. (3.3). By requiring that under permutations the wave function must change by no more than a phase factor, one actually *postulates* that the representation of the permutation group, to which the wave function belongs, is one-dimensional. Hence, the proof is based on the initial statement, which is proved then as a final result.

The second incorrectness in the proof discussed above follows from physical considerations. This proof is directly related to the behavior of the wave function. However, since the wave function is not an observable, the indistinguishability principle is related to it only indirectly via the expressions of measurable quantities. Since in quantum mechanics, the physical quantities are expressed as bilinear forms of wave functions, the indistinguishability principle requires the invariance of these bilinear forms. In Section 2.1, we represented the indistinguishability principle as the invariance of the probability density in respect to the permutations of identical particles; see also Ref. [11]:

$$|P\Psi(x_1,\ldots,x_N)|^2 = |\Psi(x_1,\ldots,x_N)|^2. \tag{3.7}$$

For systems of identical particles, the expectation values of all physical quantities described by some Hermitian operators $\hat{L}$ must also be invariant in respect to the permutations of identical particles. The operators $\hat{L}$ commute with particle permutations,

$$[P,\hat{L}] = 0. \tag{3.8}$$

Thus, the indistinguishability principle can be formulated as follows:

for identical particle systems all experimentally observable physical quantities must be permutation invariant.

It is evident that for a function to satisfy Eq. (3.7), it is sufficient that under permutations it would change as

$$P\Psi(x_1,\ldots,x_N)=e^{i\alpha_P(x_1,\ldots,x_N)}\Psi(x_1,\ldots,x_N), \tag{3.9}$$

that is, unlike the requirement of condition (3.2), in the general case the phase is a function of coordinates and the permutation, and Eq. (3.2) evidently does not hold.

I would like to mention that in an interesting paper by Lyuboshitz and Podgoretskii [22], where they discussed the cases of the indistinguishability of nonidentical particles (we discuss a similar problem in Section 2.1 on the example of protons and neutrons in nuclei), the authors presented the proof of the Pauli exclusion principle with the same errors as we discussed above. The only difference is that they started with some superposition of the wave functions and included the isotopic-spin part. It seems that the wrong proof presented by Eqs. (3.2)–(3.4) hypnotizes physicists by its simplicity.

Most proofs of the symmetry postulate contain unjustified constraints, see Refs. [10, 11, 15]. Proofs of the symmetry postulate without imposing additional constraints have been given by Girardeau [11, 23], who based it on Eq. (3.7), and in my paper [15] where it was based on the invariance of the expectation value of some one-electron operator. As was noted then by the author [16, 17, 24], the proofs, basing on the indistinguishability principle in the form (3.7) or an equivalent operator form, are also incorrect, because those equations are valid only for non-degenerate states. In a degenerate state, the system can be described with the equal probability by any one of the basis vectors of the degenerate state. As a result, we can no longer select a pure state (the one that is described by the wave function) and should regard a degenerate state as a mixed one, where each basis vector enters with the same probability. Hence, we must sum both sides of Eq. (3.7) over all wave functions which belong to the degenerate state. For instance, the probability density, which is described via the diagonal element of the density matrix, in the case of a degenerate state is expressed as

$$D_t^{[\lambda]}(x_1,\ldots,x_N;x_1,\ldots,x_N)=\frac{1}{f_\lambda}\sum_{r=1}^{f_\lambda}\Psi_{rt}^{[\lambda]}(x_1,\ldots,x_N)^*\Psi_{rt}^{[\lambda]}(x_1,\ldots,x_N), \tag{3.10}$$

where the expression (3.10) is written for the case of the f_λ-dimensional representation $\Gamma^{[\lambda]}$ of the permutation group π_N and wave functions $\Psi_{rt}^{[\lambda]}$ are constructed by the Young operators $\omega_{rt}^{[\lambda]}$, see Eq. (B.33) in Appendix B. The possibility of expressing the density matrix through only one of the functions implies that the degeneracy with respect to permutations has been eliminated. However, the latter cannot be achieved without violating the identity of the particles, and in this case the permutation group ceases to be group symmetry for the Hamiltonian.

Recently, S. Zagoulaev (St. Petersburg University, private communication) informed me that in 1937, V. Fock presented a proof of the Pauli exclusion principle in his unpublished lectures on quantum mechanics [25]. In his proof, Fock substituted the correct expression (3.9) in equation for an arbitrary operator and applied the variational theorem. However, Eq. (3.9) follows from Eq. (3.7) and the latter, as we discussed above, is valid only in the case of non-degenerate states. The Fock proof fails, if one applies it to the expression (3.10) which is valid for degenerate states.

It is not difficult to prove that for every representation $\Gamma^{[\lambda]}$ of the permutation group π_N, the probability density, Eq. (3.10), is a group invariant. In the case of an arbitrary group, it was proved in Ref. [26]. Below, I represent this proof in the case of the permutation group.

Let us apply permutation $P \subset \pi_N$ to the expression (3.10),

$$
\begin{aligned}
PD_t^{[\lambda]} &= \frac{1}{f_\lambda}\sum_r \left[\sum_u \Gamma_{ur}^{[\lambda]}(P)^* \Psi_{ut}^{[\lambda]^*} \sum_{u'} \Gamma_{u'r}^{[\lambda]}(P)\Psi_{u't}^{[\lambda]}\right] \\
&= \frac{1}{f_\lambda}\sum_{u,u'}\left(\sum_r \Gamma_{ur}^{[\lambda]}(P)\Gamma_{u'r}^{[\lambda]}(P)\right)\Psi_{ut}^{[\lambda]^*}\Psi_{u't}^{[\lambda]}.
\end{aligned}
$$

Due to the orthogonality relations for the matrix elements of irreducible representations, the sum over r is equal to $\delta_{uu'}$, and we arrive at the final result:

$$
PD_t^{[\lambda]} = \frac{1}{f_\lambda}\sum_u \left|\Psi_{ut}^{[\lambda]}\right|^2 = D_t^{[\lambda]}. \tag{3.11}
$$

This means that for all irreducible representations, characterizing the quantum states, the diagonal element of the full density matrix (and all reduced densities matrices as well) transforms according to the totally symmetric one-dimensional representation of π_N, and in this respect one cannot distinguish between degenerate and nondegenerate states. Thus, the diagonal element of the density matrix is a group invariant.

Hence, the probability density satisfies the indistinguishability principle even in the case of multidimensional representations of the permutation group. Thus, the indistinguishability principle is insensitive to the symmetry of wave function and cannot be used as a criterion for selecting the correct symmetry.

It is important to note that from the discussion above it does not follow that the symmetry of wave function is not significant, and one can perform quantum mechanical study using only the density matrix, which, as we have shown, does not depend upon the symmetry of wave function. In reality, the symmetry of wave function controls the atomic and molecular states allowed by the Pauli exclusion principle, see book [27]. We demonstrated it earlier for the rotational states of the $^{16}O_2$ molecule. Another example is the allowed multiplets in atomic spectroscopy. For instance, in the $(np)^2$ electronic shell, only ^1S, ^3P, and ^1D states are realized. The methods for finding atomic and molecular multiplets allowed by the Pauli exclusion principle are discussed in detail in Chapter 4.

Let us also stress the importance in some physical problems of the geometrical phase of the wave function studied by Berry [28]. The Berry phase is a geometrical one, which appears in addition to the familiar dynamical phase in the wave function of a quantum system that has undergone a cyclic adiabatic change. Even though in the initial and final states the system will be the same, the phases of initial and final wave functions will be different [28, 29]. As was shown by Berry [28], the Aharonov–Bohm effect [30] can be explained as a special case of the geometrical phase factor (see also Refs. [31, 32]).

Thus, the widespread application of the density functional theory (DFT) based on the Kohn–Sham equation does not mean that the wave function in quantum mechanics lost its significance. It is also important to stress that in general the concept of spin cannot be defined in the frame of the functional density formalism. McWeeny [33] formulated it in the following manner: "electron spin is in a certain sense extraneous to the DFT." As was rigorously proved by the author [26, 34], the electron density of an N-electron system does not depend on its total spin S; therefore, the conventional Kohn–Sham equations are the same for states with different S. In Ref. [26], the analysis of the existing DFT procedures developed for the study of spin–multiplet structure has been performed. It was shown that all these procedures modify only the expression for the exchange energy and use the correlation functionals (if they use them) not corresponding to the total spin of the state; see also an appropriate section in my later publication [35].

Although the Pauli exclusion principle cannot be rigorously derived from other quantum mechanical postulates, there are some heuristic arguments [16–18], indicating that the description of an identical particle system by the multidimensional representations of the permutation group leads to some contradictions with the concept of the particle identity and their independency. Hence, the experimental fact that in Nature only the one-dimensional representations of the permutation group are allowed is not accidental. In next section, we discuss these arguments in detail.

3.2 Some Contradictions with the Concept of Particle Identity and their Independence in the Case of the Multidimensional Permutation Representations

Let us consider a quantum system with an arbitrary number of identical elementary particles without the restrictions imposed by the Pauli exclusion principle. The states of a system of identical particles with the number of particles not conserved can be presented as vectors in the Fock space **F** [36]. The latter is a direct sum of spaces $\mathbf{F}^{(N)}$ corresponding to a fixed number N of particles,

$$\mathbf{F} \doteq \sum_{N=0}^{\infty} \mathbf{F}^{(N)}. \tag{3.12}$$

Each of the space $\mathbf{F}^{(N)}$ can be presented as a direct product of one-particle spaces **f**:

$$\mathbf{F}^{(N)} = \underbrace{\mathbf{f} \otimes \mathbf{f} \otimes \cdots \otimes \mathbf{f}}_{N} \tag{3.13}$$

The basis vectors of $\mathbf{F}^{(N)}$ are the product of one-particle vectors $|\upsilon_k(k)\rangle$ belonging to spaces **f**, where the argument k denotes the set of particle spin and space coordinates,

$$\left|\xi^{(N)}\right\rangle = |\nu_1(1)\rangle |\nu_2(2)\rangle \ldots |\nu_N(N)\rangle. \tag{3.14}$$

For simplicity, let us consider the case where all one-particle vectors in Eq. (3.14) are different. There will be no qualitative changes in the results, if some of the vectors coincide. Let us mention that vectors $|\upsilon_k(k)\rangle$ are the spin-orbitals, on which the total wave function is constructed; see Eq. (2.20).

One can produce $N!$ new many-particle vectors by applying to the many-particle vector (3.14) $N!$ permutations of the particle coordinates. These new vectors also belong to $\mathbf{F}^{(N)}$ and form in it a certain invariant, reducible subspace. The $N!$ basic vectors of the latter, $P|\xi^{(N)}\rangle$, make up the regular representation of the permutation group $\boldsymbol{\pi}_N$. As is known in the group theory, the regular representation is decomposed into irreducible representations, each of which appears a number of times equal to its dimension. The space $\varepsilon^{(N)}$ falls into the direct sum,

$$\varepsilon_{\xi}^{(N)} \doteq \sum_{\lambda_N} f_{\lambda_N} \varepsilon_{\xi}^{[\lambda_N]}, \tag{3.15}$$

where $\varepsilon_{\xi}^{[\lambda_N]}$ is an irreducible subspace of dimension f_λ drawn over the basis vectors $|[\lambda_N]rt\rangle$ and $[\lambda_N]$ is a Young diagram with N boxes. The basis vectors $|[\lambda_N]rt\rangle$ can be constructed of non-symmetrized basis vector $|\xi^{(N)}\rangle$ by using the Young operators $\omega_{rt}^{[\lambda_N]}$, see Appendix B, Eq. (B.33),

$$|[\lambda_N]rt\rangle = \omega_{rt}^{[\lambda_N]}\left|\xi^N\right\rangle = \left(\frac{f_\lambda}{N!}\right)^{\frac{1}{2}} \sum_P \Gamma_{rt}^{[\lambda_N]}(P) P \left|\xi^{(N)}\right\rangle, \tag{3.16}$$

where $\Gamma_{rt}^{[\lambda_N]}(P)$ are the matrix elements of representation $\Gamma^{[\lambda_N]}$ and index t distinguishes between the bases in accordance with the decomposition of $\varepsilon_{\xi}^{(N)}$ into f_λ invariant subspaces and describes the symmetry under permutations of the particle vector indices.

Thus, a space with a fixed number of particles can always be divided into irreducible subspaces $\varepsilon_{\xi}^{[\lambda_N]}$, each of which is characterized by a certain permutation symmetry given by a Young diagram with N boxes. The symmetry postulate demands that the basis vectors of a system of N identical particles must belong only to one of the two subspaces characterized by irreducible one-dimensional representations, either $[N]$ or $[1^N]$. All other subspaces are "empty." Let us examine the situation that arises when no symmetry constraints are imposed and consider the system of N identical particles described by basic vectors belonging to some irreducible subspace $\varepsilon_{\xi}^{[\lambda_N]}$.

One of the consequences of the different permutation symmetry of state vectors for bosons and fermions is the dependence of the energy of the systems on the particle statistics. For the same law of dynamic interaction, the so-called exchange terms enter the expression for the energy of fermion and boson system with opposite signs. Let us calculate the energy of a multidimensional permutation state $|[\lambda]rt\rangle$. The energy of the system in a degenerate state is

$$E = Tr(HD), \tag{3.17}$$

where D is the density operator defined, similar to Eq. (3.10), as

$$D_t = \frac{1}{f_\lambda}\sum_{r=1}^{f_\lambda} |[\lambda]rt\rangle\langle[\lambda]rt|. \tag{3.18}$$

The trace (3.17) with the density operator (3.18) is equal to the sum

$$E_t^{[\lambda]} = \frac{1}{f_\lambda}\sum_{r=1}^{f_\lambda} \langle[\lambda]rt|H|[\lambda]rt\rangle. \tag{3.19}$$

The matrix element in Eq. (3.19) has been calculated in Ref. [37] in a general case of non-orthogonal one-particle vectors. In the case where all vectors in Eq. (3.14) are different and orthogonal, one gets

$$E_t^{[\lambda]} = \sum_a \langle v_a|h|v_a\rangle + \sum_{a<b}\left[\langle v_a v_b|g|v_a v_b\rangle + \Gamma_{tt}^{[\lambda]}(P_{ab})\langle v_a v_b|g|v_b v_a\rangle\right], \tag{3.20}$$

where $\Gamma_{tt}^{[\lambda]}$ is the diagonal matrix element of the transposition P_{ab} of the arguments of vectors $|v_a\rangle$ and $|v_b\rangle$ on the right-hand side of Eq. (3.14); h and g are one- and two-particle interaction operators, respectively. Only exchange terms in Eq. (3.20) depend upon the symmetry of the state. For one-dimensional representations, $\Gamma_{tt}^{[\lambda]}(P_{ab})$ does not depend on the number of particles and the permutation P_{ab}: $\Gamma^{[N]}(P_{ab}) = 1$ and $\Gamma^{[1^N]}(P_{ab}) = -1$ for all P_{ab} and N. For multidimensional representations, the matrix elements $\Gamma_{tt}^{[\lambda]}(P_{ab})$ depend on $[\lambda]$ and P_{ab}; in general, they are different for different pairs of identical particles.[2]

Taking into account that the transitions between states with different symmetry $[\lambda]$ are strictly forbidden and each state of N particle system with different $[\lambda]$ has a different expression for its energy (they differ by the matrix element $\Gamma_{tt}^{[\lambda]}(P_{ab})$), we must conclude:

> *each type of symmetry $[\lambda_N]$ corresponds to a certain kind of particles with statistics determined by this permutation symmetry $[\lambda_N]$.*

On the other hand, the classification of state with respect to the Young diagrams $[\lambda]$ is connected exclusively with identity of particles. Therefore, it must be some additional *inherent particle characteristics,* which establishes for the N particle system to be in a state with definite permutation symmetry (like integer and half-integer values of particle spin for bosons and fermions), and this inherent characteristic must be different for different $[\lambda_N]$. Hence, the particles belonging to the different types of permutation symmetry $[\lambda_N]$ are not identical.

Let us trace down the genealogy of irreducible subspaces $\varepsilon_{\xi}^{[\lambda_N]}$. In Fig. 3.1, the genealogy for all irreducible subspaces with $N = 2$–4 is presented. We called the hypothetical particles characterized by the multidimensional representations of the permutation group as *intermedions,* implying that they obey some intermediate between fermion and boson statistics.

For bosons and fermions, there are two nonintersecting chains of irreducible representations: $[N] \to [N+1]$ and $\left[1^N\right] \to \left[1^{N+1}\right]$ for all N; the energy expression for each type of particles has the same analytical form that does not depend on N. The situation changes drastically, if we put into consideration the multidimensional representations. The number of different statistics depends upon the number of particles in a system and rapidly increases with N. For the multidimensional representations, we cannot select any nonintersecting chains, as in the fermion and boson cases. According to Fig. 3.1, the intermedion particles with a definite $[\lambda_N]$ in the Nth generation can originate from particles of different kinds $[\lambda_{N-1}]$ in the $(N-1)$th generation, even from fermions or bosons. Thus, if we reduce the state of N-particle system described by some symmetry $[\lambda_N]$ on one particle, the particles in the $(N-1)$th generation must in general be described by a linear

[2] The matrices of transpositions for all irreducible representations of groups $\pi_2 - \pi_6$ are presented in book [27], appendix 5.

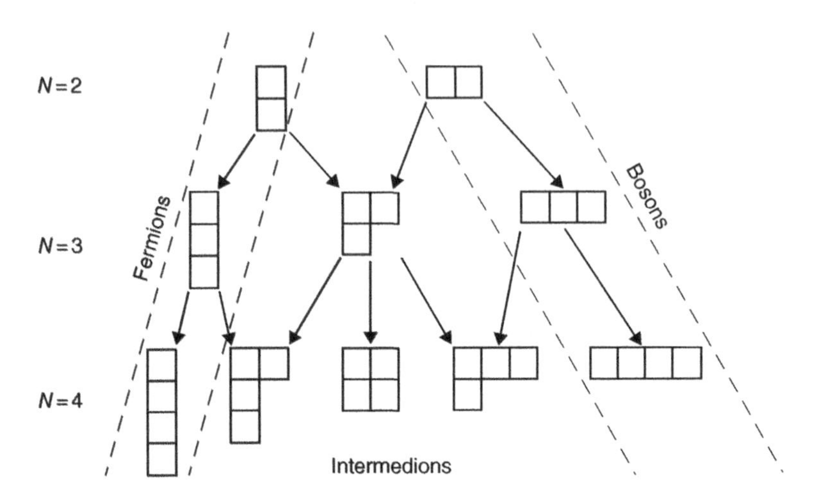

Fig. 3.1 The Young diagrams for $N = 2-4$ and their genealogy

combination of wave functions with different permutations symmetry $[\lambda_{N-1}]$. However, this linear combination does not describe identical particles; see Eq. (3.26) and its discussion.

The physical picture in which adding one particle changes properties of all particles cannot correspond to a system of *independent* particles (although, it cannot be excluded for some quasiparticle systems where quasiparticles are not independent, see Chapter 5). For an ideal gas, it is evident that adding a particle identical to a system of N identical particles cannot change the properties of a new $(N+1)$-particle system. On the other hand, the interaction of identical particles does not change the permutation symmetry of noninteracting particle system. This can be rigorously proved [15]. Namely:

The wave vector $|\Psi\rangle$ of a system of interacting particles characterized by the Hamiltonian

$$H = H_0 + V, \tag{3.21}$$

where H_0 is the Hamiltonian of noninteracting system and V is an interaction operator, defined in the same Fock space as the wave vector $|\Psi_0\rangle$ of noninteracting system, and it can be generated from the latter by using some unitary transformation,

$$|\Psi\rangle = \hat{U}|\Psi_0\rangle. \tag{3.22}$$

The form of this unitary transformation can be obtained if we use the well-known relationship that the Brillouin–Wigner perturbation theory [38] is based on:

$$|\Psi\rangle = |\Psi_0\rangle + \frac{Q}{E - H_0} V|\Psi\rangle, \tag{3.23}$$

where Q is the projection operator that projects an arbitrary vector onto the multitude of vectors of Hilbert (Fock) space that are orthogonal to the vector $|\Psi_0\rangle$. Upon substituting Eq. (3.22) into Eq. (3.23), we get an equation for $\hat{U}$, from which follows

$$\hat{U}^{-1} = 1 - \frac{Q}{E - H_0} V. \tag{3.24}$$

Since the interaction operator V of identical particles is invariant with respect to permutations of identical particles, the operator $\hat{U}$ is also invariant. Hence, according to Eq. (3.22), the states $|\Psi\rangle$ and $|\Psi_0\rangle$ have the same permutation symmetry.

Thus, the scenario, in which all symmetry types $[\lambda_N]$ are allowed and each of them corresponds to a definite particles statistics, contradicts to the concept of particle identity and their independency from each other.

Let us consider the possibility that for some type of intermedions with fixed N, a multidimensional representation $[\lambda_N]$ exists that stems only from one $[\lambda_{N-1}]$, as $[2^2]$ in the case $N = 4$ (see Fig. 3.1). But in the process of reducing the number of particles, we cannot avoid the case $N = 3$ where only one multidimensional representation exists with $[\lambda_3] = [21]$, and this representation proceeds from both two-particle representations: $[\lambda_2] = [2]$, corresponding to bosons, and $[\lambda_2] = [1^2]$, corresponding to fermions. However, contrary to the statement in Refs. [39, 40], where hypothetical paraparticles were considered, the wave function of two identical particles cannot be described by some superposition.

$$\Psi_n(x_1, x_2) = c_1 \Psi^{[2]}(x_1, x_2) + c_2 \Psi^{[1^2]}(x_1, x_2). \tag{3.25}$$

This superposition corresponds to nonidentical particles, since it does not satisfy the indistinguishability principle. In fact,

$$P_{12} |\Psi_n(x_1, x_2)|^2 = \left| c_1 \Psi^{[2]}(x_1, x_2) - c_2 \Psi^{[1^2]}(x_1, x_2) \right|^2 \neq |\Psi_n(x_1, x_2)|^2. \tag{3.26}$$

Let us stress that the permutation group can be applied only to identical particles, and these particles are transformed according to the irreducible representations $\Gamma^{[\lambda]}$ of the permutation group, but not according to their linear combinations.

The situation labeled in Refs. [39, 40] as “disagreeable” or “unpalatable” is physically forbidden, if we consider a system of identical particles. The two identical particles must be considered only in the pure fermion or boson state. However, if there are no prohibitions on the permutation symmetry, the addition of the third particle identical to the two others can change the fermion (or boson) statistics on the intermedion statistics with $[\lambda_3] = [21]$, and this takes place even in an ideal gas of intermedions. Again, we obtained a contradiction with the concept of particle identity and their independence.

It is worthwhile to mention that the multidimensional representations of the permutation group can be used in quantum mechanics of identical particles, although not for the total wave function but for its factorized parts. In Section 2.3.2, we discuss the case of a spinless Hamiltonian when the total spin S is a good quantum number and labels the energy levels of the system. In this case, the total wave function can be presented as a product of spatial and spin wave functions, symmetrized according to the conjugated irreducible representations of the permutation group with the dual Young diagrams. On the one hand, this provides the condition of the antisymmetry of the total wave function. On the other hand, the symmetry of the space wave functions depends on the value of the total spin S, which causes the dependence on S of the total energy of the system regardless of the dynamical interaction law. Hence, the system of fermions can be described by the multidimensional representations of the permutation group; however, they correspond not to the total wave function but to its parts. The total wave function is antisymmetric in accordance with the Pauli exclusion principle.

As was demonstrated in this section, the permission of multidimensional representations of the permutation group for the total wave function leads to contradictions with the concepts of particle identity and their independence. All contradictions in discussed scenarios are resolved, if only the one-dimensional irreducible representations of the permutation group are permitted. Thus, although the Pauli exclusion principle cannot be derived from other fundamental principles of quantum mechanics, it may not be considered as a postulate, since all symmetry options for the total wave function, except the one-dimensional irreducible representations, cannot be realized. These arguments can be considered as a basis for the existence of the Pauli exclusion principle. Nevertheless, the problem why the connection between the value of spin and the permutation symmetry of wave function exists (spin–statistics connection) is still unsolved.

References

[1] I. Duck and E.C.G. Sudarshan, *Pauli and the Spin-Statistics Theorem*, World Scientific, Singapore, 1997.
[2] I. Duck and E.C.G. Sudarshan, *Am. J. Phys.* **66**, 284 (1998).
[3] E.C.G. Sudarshan, *Stat. Phys. Suppl.: J. Indian. Inst. Sci.*, June, 123, (1975).
[4] A.S. Wightman, *Am. J. Phys.* **67**, 742 (1999).
[5] In *Feynman Lectures on Physics*, Vol. **III**, R.P. Feynman, R.B. Leighton, and M. Sands, Addison-Wesley, Reading, Massachusetts, 1965, p. 3.
[6] W. Pauli, *Science* **103**, 213–215 (1946).
[7] P.A.M. Dirac, *The Principles of Quantum Mechanics*, Clarendon Press, Oxford, 1958.
[8] L.I. Schiff, *Quantum Mechanics*, McGraw-Hill, New York, 1955.
[9] A.M. Messiah, *Quantum Mechanics*, North-Holland, Amsterdam, 1962.
[10] A.M. Messiah and O.W. Greenberg, *Phys. Rev.* **136**, B248 (1964).
[11] M.D. Girardeau, *Phys. Rev.* **139**, B500 (1965).

[12] E.M. Corson, *Perturbation Methods in Quantum Mechanics of Electron Systems*, University Press, Glasgow, 1951.
[13] L.D. Landau and E.M. Lifshitz, *Quantum Mechanics, (Nonrelativistic Theory)*, 3rd edn, Pergamon Press, Oxford, 1977.
[14] Blokhintzev, *Principles of Quantum Mechanics*, Allyn and Bacon, Boston, Massachusetts, 1964.
[15] I.G. Kaplan, *Sov. Phys. Uspekhi* **18**, 988 (1976).
[16] I.G. Kaplan, *J. Mol. Struct.* **272**, 187 (1992).
[17] I.G. Kaplan, *Int. J. Quantum Chem.* **89**, 268 (2002).
[18] I.G. Kaplan, *Found. Phys.* **43**, 1233 (2013).
[19] G.S. Canright and S.M. Girvin, *Science*, **247**, 1197 (1990).
[20] R. Mirman, *Nuovo Cimento* **18B**, 110, 1973.
[21] L. Piela, *Ideas of Quantum Chemistry*, 2nd edn, Elsevier, Amsterdam, 2014.
[22] V.L. Lyuboshitz and M.I. Podgoretskii, *Sov. Phys. - JETP* **28**, 469 (1969); **30**, 100 (1970).
[23] M.D. Girardeau, *J. Math. Phys.* **10**, 1302 (1969).
[24] I.G. Kaplan, in *Group Theoretical Methods in Physics*, Vol. **1**, ed. V.I. Man'ko, Nauka, Moscow, 1980, pp. 175–181.
[25] V.A. Fock, *Lectures on Quantum Mechanics*. Manuscript, B31, F-750, Library of Physical Faculty, St Petersburg State University, 1937.
[26] I.G. Kaplan, *Int. J. Quantum Chem.* **107**, 2595 (2007).
[27] I.G. Kaplan, *Symmetry of Many-Electron Systems*, Academic Press, New York, 1975.
[28] M.V. Berry, *Proc. Roy. Soc. Lond. A* **392**, 45 (1984).
[29] M.V. Berry, Scientific American, December, 26, 1988.
[30] Y. Aharonov and D. Bohm, *Phys. Rev.* **115**, 485 (1959).
[31] C.A. Mead and D.G. Truhlar, *J. Chem. Phys.* **70**, 2284, 1979.
[32] C.A. Mead, *Chem. Phys.* **49**, 23; **49**, 33 (1980).
[33] R. McWeeny, *Philos. Mag. B* **69**, 727 (1994).
[34] I.G. Kaplan, *J. Mol. Struct.* **838**, 39 (2007).
[35] I.G. Kaplan, *Int. J. Quantum Chem.* **112**, 2858 (2012).
[36] S.S. Schweber, *An Introduction to Relativistic Quantum Field Theory*, Row Peterson, New York, 1961.
[37] I.G. Kaplan and O.B. Rodimova, *Int. J. Quantum Chem.* **7**, 1203 (1973).
[38] N.H. March, W.H. Young, and S. Sampanthar, *The Many-Body Problem in Quantum Mechanics*, Cambridge University Press, Cambridge, 1967.
[39] A. Casher, G. Frieder, M. Gluck, and A. Peres, *Nucl. Phys.* **66**, 632 (1965).
[40] J.B. Hartle and J.R. Taylor, *Phys. Rev.* **178**, 2043 (1969).

4

Classification of the Pauli-Allowed States in Atoms and Molecules

4.1 Electrons in a Central Field

4.1.1 Equivalent Electrons: L–S *Coupling*

In this chapter I describe methods developed for finding the Pauli-allowed states for atoms and molecules. It is reasonable to start from atoms, really from electronic shells in atoms. For classification of molecular states often we need to know the permitted states of atoms that constitute the molecule. The application of group theory allows to create very elegant and clear methods for finding the terms permitted by the Pauli exclusion principle. In Chapter 2 we discussed the application of the permutation group for constructing the total wave function of a system. In the case when the spatial coordinates and spin variables may be separated, the total wave function can be presented as an antisymmetric product of a spatial wave function and a spin wave function, which described the state with the definite total spin S. As we show later, this approach is very useful for finding the Pauli-allowed atomic and molecular states. In these methods it is also useful to apply the linear groups and their interconnection with the permutation group. The necessary mathematical apparatus is represented in Appendix C.

In quantum calculations of many-electron systems the so-called one-electron approximation is widely used. This approximation is a starting point of the Hartree–Fock method, which is, in its turn, a starting point of the most modern molecular computational methods, in which the electron correlation is taken into account. In the Hartree–Fock method each electron is regarded as being in a stationary state in the field of the nuclei and of all the other electrons. For atoms this

The Pauli Exclusion Principle: Origin, Verifications, and Applications, First Edition. Ilya G. Kaplan.

field can be taken to be centrally symmetric to a very good approximation, and the one-electron states can be classified according to the irreducible representations of the orthogonal group $\mathbf{O_3}$. The latter can be represented as a direct product of groups:

$$\mathbf{O_3} = \mathbf{C_i} \times \mathbf{R_3},$$

where $\mathbf{C_i}$ consists of two elements: unit element E and inversion I, and $\mathbf{R}_3$ is the rotation group in the three-dimensional space. Beside the parity, which is a consequence of the symmetry group $\mathbf{C_i}$, the state of each electron is characterized by a set of four quantum numbers: n, l, m, and σ, where n is the principal quantum number, l is the orbital angular momentum, m is the projection of l upon the z-axis, and σ is the projection of the spin of the electron upon the z-axis. In determining the energy levels of atoms with low atomic weights, one usually neglects relativistic interactions. As a result, the spin and orbital states of a system can be considered separately. This approximation is known as *Russell–Saunders* or *L–S coupling*.

In the L–S coupling approximation the total spin of the electrons S and the total orbital angular momentum L are conserved. Values of L and S characterize the energy levels of the system. In addition, each energy level is characterized by the configuration of one-electron states from which it arises. In order to specify an electronic configuration, it suffices to assign the principal quantum number n and the orbital angular momentum l for each electron. Electrons with the same n and l quantum numbers are said to be *equivalent*. A set of equivalent electrons of a given type constitutes a *shell*. Since, according to the Pauli exclusion principle, each electron must be characterized by a distinct set of four quantum numbers n, l, m, and σ, a shell $(nl)^N$ may contain no more than $2(2l+1)$ electrons. A shell, in which all the states are filled, is said to be *closed*.

The total orbital and spin angular momenta are obtained by the vector coupling of angular momenta of the individual electrons. The coupling of angular momenta can be conveniently represented in the form of a diagram. Figure 4.1 shows the diagram for $l=1$. The numbers enclosed in circles give the number of times the given value for the total angular momentum L occurs in a system of N electrons.[1] We denote this number by a_{LN}. Since the number of states for a system of N angular momenta is unchanged by the coupling,

$$\sum_L a_{LN}(2L+1) = (2l+1)^N.$$

This equation is easily verified by Fig. 4.1. Thus for $N=4$ we have

[1] The diagram corresponds in fact to the decomposition of the direct product of N irreducible representations $D^{(l)}$ of the group $\mathbf{R}_3$ into irreducible representations $D^{(L)}$.

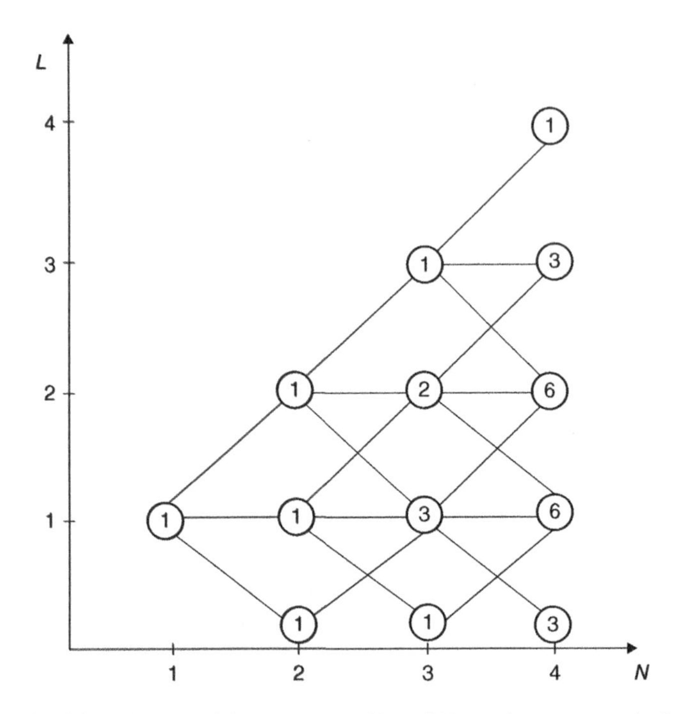

Fig. 4.1 Diagram of the vector coupling of N angular momenta $l = 1$

$$3 \times 1 + 6 \times 3 + 6 \times 5 + 3 \times 7 + 1 \times 9 = 81.$$

The a_{LN}, however, are by no means equal to the number of different energy levels of a configuration $(nl)^N$ with the given value of L. In fact, a part of states with the same L differ from one another only by a permutation of the electrons, and because of the indistinguishability of electrons, such states must possess the same energy (this is known as the *permutational degeneracy*). Some of the remaining states are forbidden, since they do not satisfy the Pauli exclusion principle. The true number of energy levels for a system of equivalent electrons can be quite simple determined from group-theoretical considerations, see below.

In the one-particle approximation, the spatial wave function for an electronic configuration $(nl)^N$ is constructed, if one neglects the electronic interactions, as a product of N one-electron functions, usually named *orbitals*:

$$\Phi_0 = \varphi_{nlm_1}(1)\varphi_{nlm_2}(2)\ldots\varphi_{nlm_N}(N). \tag{4.1}$$

In Eq. (4.1), the number i in the argument of an orbital denotes the set of coordinates for the ith electron. One can form $(2l+1)^N$ such products altogether, all of which in the absence of interactions between the electrons belong to a single energy level. Orbitals φ_{nlm} which differ in the quantum number m can be regarded as basis vectors for a $(2l+1)$-dimensional vector space. Under unitary transformations of this space the products (4.1) transform as tensors and generate a basis for a $(2l+1)^N$-dimensional tensor representation of the unitary group $\mathbf{U}_{2l+1}$. The decomposition of such representation is discussed in Appendix C, Section C.3.2. From the results of that section it follows that the number of times, with which each irreducible representation $U_{2l+1}^{[\lambda]}$ occurs in the decomposition of the tensor representation, is equal to the dimension f_λ of the irreducible representation of the permutation group $\Gamma^{[\lambda]}$ characterized by the same Young diagram $[\lambda]$, in which the number of cells in any column does not exceed $2l+1$. This means that there always exists a linear transformation which transforms the initial $(2l+1)^N$ functions (4.1) into sets of functions each of which is characterized by a specified Young diagram of N cells whose columns do not exceed $2l+1$ in length. Such a set contains $f_\lambda \delta_\lambda(2l+1)$ functions, where $\delta_\lambda(2l+1)$ is the dimension of the representation $U_{2l+1}^{[\lambda]}$. The functions in each set can be represented as a point on a plane diagram placed in the form of a rectangle:

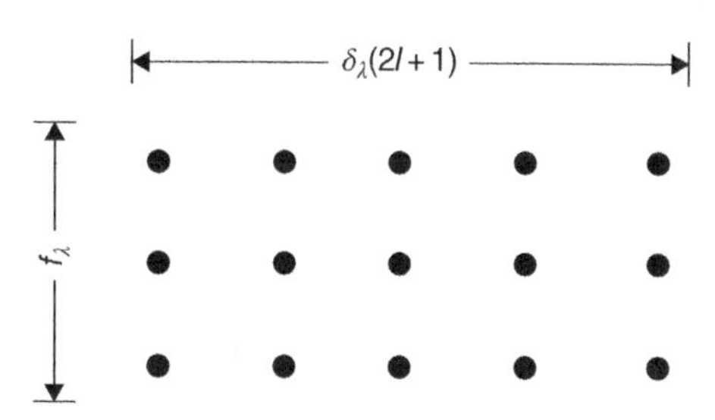

Functions, which lie on a single row of this rectangle, transform according to the irreducible representation $U_{2l+1}^{[\lambda]}$; functions lying on a single column transform according to the irreducible representation $\Gamma^{[\lambda]}$ of the permutation group π_N and are enumerated by the tableaux r.

For example, for $l=1$, $N=4$ (the configuration p^4), the 81 functions (4.1) divide into four sets corresponding to $[\lambda]=[4]$, $[31]$, $[2^2]$, and $[21^2]$ ($[\lambda]=[1^4]$ is not allowed). From formulae (B.21) and (C.76) we find the dimensions of irreducible representations $\Gamma^{[\lambda]}$ and $U_3^{[\lambda]}$.

$[\lambda]$:	[4]	[31]	$[2^2]$	$[21^2]$
$\delta_\lambda(3)$:	15	15	6	3
f_λ:	1	3	2	3

It is easily verified that $\sum_\lambda f_\lambda \delta_\lambda = 81$. We thus arrive at a diagram similar to that given in Section C.3 for the case $l=1$, $N=3$.

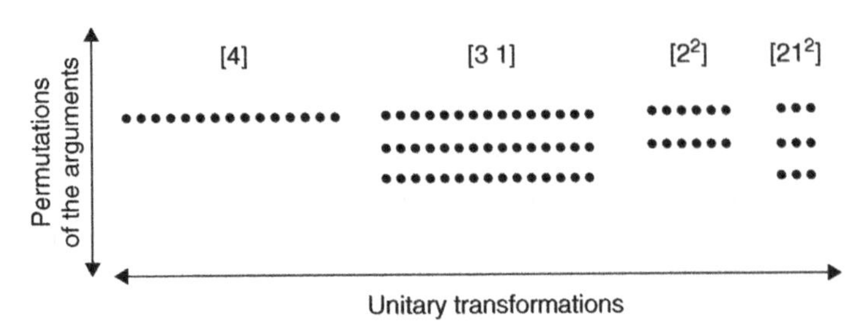

As we discussed in Section 2.3.2, in the nonrelativistic approximation, which corresponds to *L–S* coupling case, the antisymmetric total wave function $\Psi^{[1^N]}$ can be represented as a linear combination of the products of *N*-electron spatial, Φ, and spin, Ω, wave functions symmetrized according to irreducible representations $\Gamma^{[\lambda]}$ of the permutation group

$$\Psi^{[1^N]} = \frac{1}{\sqrt{f_\lambda}} \sum_r \Phi_r^{[\lambda]} \Omega_{\tilde{r}}^{[\tilde{\lambda}]}, \tag{4.2}$$

where $\Gamma^{[\tilde{\lambda}]}$ denotes the representation *conjugate* to $\Gamma^{[\lambda]}$; its matrix elements are $\Gamma_{\tilde{r}\tilde{t}}^{[\tilde{\lambda}]}(P) = (-1)^p \Gamma_{rt}^{[\lambda]}(P)$ where p is the parity of the permutations P. The Young diagram $[\tilde{\lambda}]$ is dual to $[\lambda]$, that is, it is obtained from the latter by replacing rows by columns. Since each electron possesses spin $s = 1/2$, that is, there is the one-to-one correspondence between $[\tilde{\lambda}]$ and the total spin S, the symmetrization of a spin function according to a Young diagram $[\tilde{\lambda}]$ automatically leads to the result that functions $\Omega_{\tilde{r}}^{[\tilde{\lambda}]}$ with differing values of the projection of the spin angular momentum transform according to irreducible representations $U_2^{[\lambda]}$ of the unitary group $\mathbf{U}_2$. The irreducible representations $U_2^{[\tilde{\lambda}]}$ are characterized by the Young diagrams $[\tilde{\lambda}]$, which contain no more than two rows, because the electron spin function is a two-component quantity corresponding to the two spin projections $\pm 1/2$. Since Young diagrams $[\lambda]$ and $[\tilde{\lambda}]$ are dual to each other, the $[\lambda]$ diagrams may not contain more than two columns.

From this it follows that the only orbital states, which are realized in the electron configuration considered above for $l = 1$, $N = 4$, are those that correspond to the rectangles with $[\lambda] = [2^2]$ and $[\lambda] = [21^2]$ in the diagram presented above. In addition, it should be noted that all functions that lie on a single column of a rectangle enter into one antisymmetric combination (4.2); therefore, each column corresponds to a single physical state. The number of physically distinct orbital states is determined by the number of columns, that is, is equal to $\sum_\lambda \delta_\lambda(2l+1)$. We have in this case altogether $\delta_{[2^2]}(3) + \delta_{[21^2]}(3) = 6 + 3 = 9$ permitted orbital states.

In order to determine the values of the total orbital angular momentum L that correspond to the permitted representations $U_{2l+1}^{[\lambda]}$, it is necessary to carry out the reduction $\mathbf{U}_{2l+1} \rightarrow \mathbf{R}_3$. On doing so, the irreducible representations $U_{2l+1}^{[\lambda]}$ decompose into irreducible representations $D^{(L)}$ of the group $\mathbf{R}_3$. This process has been described in detail in Appendix C, Section C.3.4. The decomposition for the configuration p^4 was given by Eqs. (C.93). From these equations it follows that the representations of the spatial wave functions correspond to the following values of L:

$$\begin{array}{lll} [\lambda]_{\text{orb}} : & [2^2] & [21^2] \\ & L=2,0 & L=1. \end{array} \tag{4.3}$$

The associated representations of the spin wavefunctions each correspond uniquely to a single value of the total spin (this value can be easily found from the form of the Young diagram $[\tilde{\lambda}]$ by Eq. (2.56)) in Chapter 2:

$$\begin{array}{lll} [\lambda]_{\text{spin}} : & [2^2] & [31] \\ & S=0 & S=1. \end{array}$$

Consequently, a configuration of four equivalent electrons, p^4, gives rise to the following permitted multiplets[2]:

$$^{2S+1}L:\ {}^1S, {}^3P, \text{ and} {}^1D.$$

The configuration p^2, which is the complementary configuration to a closed shell, gives rise to the same multiplets. This result follows from the equivalent relation (C.86).

Thus, in order to determine the allowed multiplets of a configuration of equivalent electrons $(nl)^N$ we propose the following general procedure: all the allowed Young diagrams with N cells that consist of two columns whose lengths do not exceed $2l+1$ are first written out and put into correspondence with the appropriate dual spin diagrams. The permitted values of L are then found from the reduction $\mathbf{U}_{2l+1} \rightarrow \mathbf{R}_3$ (which in practice can be taken from the tables in Appendix C,

---

[2] Values of the total orbital angular momentum L for atomic terms are denoted by capital letters:

	S	P	D	F	G	H	I	K
For $L=$	0	1	2	3	4	5	6	7

The letter S for $L=0$ should not be confused with the notation S for the total spin of a system.

Section C.4), and the multiplicity derived from the form of the Young spin diagram. This procedure can be represented schematically as follows[3]:

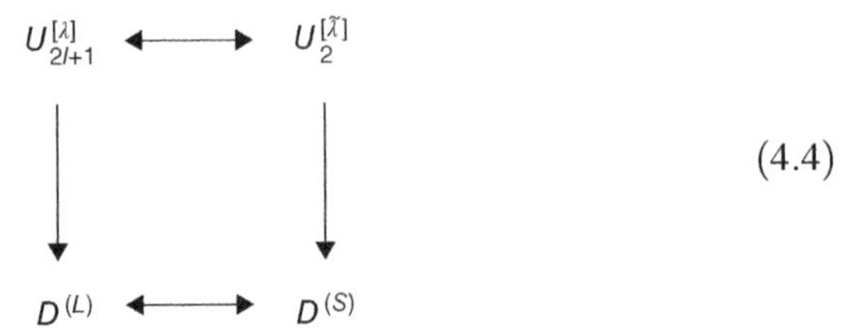

(4.4)

The simplest example is that of the two-electron configuration $(nl)^2$. Study of the reduction $\mathbf{U}_{2j+1} \rightarrow \mathbf{R}_3$ in Section C.3 showed that in the decomposition of $U^{[2]}_{2l+1}$ only representations $D^{(L)}$ with L even may occur and the decomposition of $U^{[1^2]}_{2l+1}$ contains only representations $D^{(L)}$ with L odd. Since the spin representation $U^{[1^2]}_2$ corresponds to $S=0$ and $U^{[2]}_2$ to $S=1$, the application of scheme (4.4) allows one to arrive at a general conclusion:

in the electron configuration $(\mathrm{nl})^2$ *terms with even values of* L *are singlets and terms with odd values of* L *are triplets.*

As another example for the application of scheme (4.4), we determine the allowed multiplets for the configuration d^3. It is convenient to present the results of this study in the form of a table (the values of L in the first column of Table 4.1 are taken from the Table C.4 in Appendix C).

Table 4.1 Allowed terms for the configuration d^3

$U^{[\lambda]}_5 \rightarrow D^{(L)}$	$U^{[\tilde{\lambda}]}_2 \rightarrow D^{(S)}$	Multiplets
$[21]\ L=1,2^2,3,4,5$	$[21]\ S=1/2$	${}^2P$, $({}^2D)^2$, ${}^2F$, ${}^2G$, ${}^2H$
$[1^3]\ L=1,3$	$[3]\ S=3/2$	${}^4P$, ${}^4F$

[3] Tables of the permitted multiplets already exist for equivalent electrons. These have been derived by a different quite cumbersome method, which does not require the application of group theory, for example, see Ref. [1], section 67. However, the group-theoretical method given here is more transparent than the conventional method, and furthermore, it allows a more detailed classification in cases where terms occur more than once. The scheme is also of methodological interest because it can be applied in a number of more complicated situations (e.g., in finding the allowed molecular multiplets (see Sections 4.2 and 4.3)).

4.1.2 *Additional Quantum Numbers: The Seniority Number*

From the fact that a single representation $D^{(S)}$ corresponds to each representation $U_2^{[\lambda]}$ (each spin Young diagram corresponds to only one value of S) and that, because of the Pauli exclusion principle, there is a one-to-one correspondence between the orbital and spin Young diagrams, the specification of the orbital Young diagram is equivalent to specifying the total spin S. Atomic terms can therefore be enumerated by the quantum numbers L and $[\lambda]$ instead of by the quantum numbers L and S. The assignment of a Young diagram and orbital angular momentum L to each term is equivalent to classifying the orbital states by the irreducible representations of the groups U_{2l+1} and $\mathbf{R}_3$. The specification of the projection of the orbital angular momentum along the z-axis, M, signifies an additional classification of the orbital states by the irreducible representations of the two-dimensional rotation group $\mathbf{R}_2$.

A classification of this kind for a configuration p^N uniquely determines a state. However, in the case of d^N and f^N configurations several states with the same L may correspond to a single symmetry diagram $[\lambda]$. Thus, for the configuration d^3 two states with $L=2$ possess the identical permutations symmetry $[\lambda]=[21]$ (see Table 4.1). In order to distinguish between repeated terms, it is useful to introduce additional quantum numbers.[4] For this purpose one would naturally make use of some subgroup of U_{2l+1} which itself contains $\mathbf{R}_3$ as a subgroup. Racah [3] showed that the $(2l+1)$-dimensional rotation group $\mathbf{R}_{2l+1}$ is such an additional group. The spatial wavefunctions of a configuration are then constructed so as to be simultaneously basis functions for irreducible representations of the following chain of subgroups:

$$\mathbf{U}_{2l+1} \supset \mathbf{R}_{2l+1} \supset \mathbf{R}_3 \supset \mathbf{R}_2. \tag{4.5}$$

According to the definition of the group $\mathbf{R}_{2l+1}$, the operations in it leave invariant the scalar product of two vectors defined in the $(2l+1)$-dimensional space. The only scalar product that is invariant with respect to the operations of both the groups $\mathbf{R}_{2l+1}$ and $\mathbf{R}_3$ is the wavefunction of the configuration l^2 for the state $L=0$ (see Eq. (D.8) in Appendix D):

$$\psi_0^{(0)} = \frac{(-1)^l}{(2l+1)^{1/2}} \sum_m (-1)^m \psi_m^{(l)}(1)\psi_{-m}^{(l)}(2). \tag{4.6}$$

A system of two electrons with $L=0$ has a value of $S=0$ (see the rule after Eq. 4.4). Such electrons with paired spins are said to form a closed pair. If a closed pair is added to a configuration l^{N-2}, then because of the invariance with respect to

[4] As a rule such additional quantum numbers are not "good" quantum numbers, that is, the matrix of the Hamiltonian is not diagonal in them. The δ-forces [2] are a special example of interactions for which the seniority number (introduced below) is, in fact, a good quantum number.

operations of the group $\mathbf{R}_{2l+1}$, the formed configuration $l^{N-2}\ l^2(^1S)$ belongs to the same irreducible representation of $\mathbf{R}_{2l+1}$ as the configuration l^{N-2}. This circumstance allows each irreducible representation of the group $\mathbf{R}_{2l+1}$ to be characterized by the minimum number of electrons v of a configuration l^v for which the particular representation first makes its appearance. In addition, it is obvious that terms whose L values match the L structure of the irreducible representation of $\mathbf{R}_{2l+1}$ corresponding to a given v, make their first appearance in the configuration l^v. Each term can therefore be characterized by the additional quantum number v. This number is called the *seniority number*, and is usually written as a left suffix to the multiplet symbol: ${}^{2S+1}_{v}L$. Thus the two ${}^2D$ terms of the configurations d^3 are distinguished by the seniority number, and are written ${}^2_1D$ and ${}^2_3D$.

Thus, repeated multiplets are classified by values of the seniority number v signifying the configuration in which the given term makes its first appearance. The presence of a state $[\lambda]vL$ in a configuration l^N means that there are $(N - v)/2$ closed pairs in the given configurations. We emphasize that the Hamiltonian matrix is not diagonal in the number v. This is due to the fact that the Hamiltonian is not, in general, invariant with respect to the operations of $\mathbf{R}_{2l+1}$. Nevertheless, such a classification substantially simplifies both the construction of wavefunctions for repeated terms and the calculation of matrix elements (see Judd [4]).

The set of quantum numbers $[\lambda]$, v, and L uniquely characterizes the terms of a configuration d^N. For a configuration f^N it turns out that this classification is insufficient, terms that possess identical sets $[\lambda]$, v, and L occurring more than once. However, for the f-shell (and only in this case) an extra group can be added to the chain of groups (4.5) This is the group of orthogonal transformations in seven-dimensional space which leaves a trilinear antisymmetric form invariant [3, 4]. This group is usually denoted by $\mathbf{G}_2$.[5] As a result, the states, which arise from the configurations f^N, are uniquely classified by means of the irreducible representations of the extended chain of groups:

$$\mathbf{U}_7 \supset \mathbf{R}_7 \supset \mathbf{G}_2 \supset \mathbf{R}_3 \supset \mathbf{R}_2. \tag{4.7}$$

4.1.3 Equivalent Electrons: j–j Coupling

In heavy atoms it is necessary to take into account the spin–orbit interaction even in the zeroth order of perturbation theory. The one-electron energy levels are characterized by values of the total angular momentum of an electron, j, obtained by vector coupling the orbital and spin angular momenta:

[5] The classification of states according to the irreducible representations of the group $\mathbf{G}_2$ is equivalent to the diagonalization of a certain scalar operator which simultaneously belongs to irreducible representations of the groups $\mathbf{R}_7$ and $\mathbf{G}_2$ [4]. An analogous scalar operator for the group $\mathbf{R}_9$ has been introduced in Ref. [5]. The eigenvalues of this operator facilitate the classification of the electronic terms of the configuration g^N [5].

$$\mathbf{j}=\mathbf{l}+\mathbf{s}. \tag{4.8}$$

Since $s=1/2$, j takes one of the two values for a given value of l: $j=l\pm 1/2$. The wavefunctions describing the states of a single electron in this case form bases for the irreducible representations $D^{(j)}$ of the group $\mathbf{R}_3$ that occur in the decomposition of the direct product $D^{(l)}\times D^{(1/2)}$. According to the general formula (C.45) in Appendix C,

$$\psi_{nj\mu}=\sum_{m,\sigma}\varphi_{nlm}\chi_{s\sigma}\langle lm,s\sigma|j\mu\rangle, \tag{4.9}$$

where $\langle lm, s\sigma| j\mu\rangle$ are the Clebsch–Gordan coefficients.

The energy levels of an N-electron system are characterized by a resultant angular momentum J, which is obtained by vector coupling the angular momenta j of the individual electrons. This scheme for constructing the energy levels is called the *j–j coupling scheme*. In this scheme, the electrons with identical values of n and j are equivalent. Because of the Pauli exclusion principle, it is obvious that a configuration of equivalent electrons $(nj)^N$ cannot contain more than $2j+1$ electrons, while in the L–S coupling scheme the shell $(nl)^N$ can contain up to $2(2l+1)$ electrons.

In the zeroth approximation, in which the electronic interactions are neglected, the total wavefunction of a configuration $(nj)^N$ is constructed as a product of one-electron spin-orbitals (4.9):

$$\Psi_0=\psi_{nj\mu_1}(1)\psi_{nj\mu_2}(2)\ldots\psi_{nj\mu_N}(N). \tag{4.10}$$

In contrast to the L–S coupling case, in which one constructs from the product of orbitals (4.1) a spatial wavefunction $\Phi_r^{[\lambda]}$ with the permutation symmetry of a Young diagram $[\lambda]$, in j–j coupling one constructs from spin-orbitals (4.10) a total wavefunction whose permutation symmetry, according to the Pauli exclusion principle, is confined to the antisymmetric representation $[\lambda]=[1^N]$.

The values of the total angular momentum J that are allowed for a configuration $(nj)^N$ are just those for which the coupled function is antisymmetric. For the configuration $(nj)^2$ it follows from relations (C.49) in Appendix C that

$$P_{12}\psi_M^{(J)}(j(1)j(2))=-(-1)^J\psi_M^{(J)}(j(1)(2)), \tag{4.11}$$

since $(-1)^{2j}=-1$. Hence, J can only take even values due to the antisymmetry condition. In the case of an arbitrary configuration $(nj)^N$ the allowed values of J are found from the decomposition

$$U_{2j+1}^{[1^N]}\doteq\sum_J a^{(J)}D^{(J)}. \tag{4.12}$$

Table 4.2 Allowed values of the total angular momentum J for the configuration j^N

j	N	J
$\frac{1}{2}$	1	$\frac{1}{2}$
$\frac{3}{2}$	1	$\frac{3}{2}$
	2	0, 2
$\frac{5}{2}$	1	$\frac{5}{2}$
	2	0, 2, 4
	3	$\frac{3}{2}, \frac{5}{2}, \frac{9}{2}$
$\frac{7}{2}$	1	$\frac{7}{2}$
	2	0, 2, 4, 6
	3	$\frac{3}{2}, \frac{5}{2}, \frac{7}{2}, \frac{9}{2}, \frac{11}{2}, \frac{15}{2}$
	4	0, 2 (twice), 4 (twice), 5, 6, 8

The procedure for carrying out such decompositions is described in Appendix C, Section C.3.4. In Table 4.2 the allowed values of J for configurations $(nj)^N$ from $j = 1/2$ to $j = 7/2$ are represented. Only those configurations are given, for which the j shell is not more than half-filled, since the J values for the configurations $(nj)^N$ and $(nj)^{2j+1-N}$ coincide. It can be seen from this table that for the configuration $(7/2)^4$ terms with $J = 2$ and $J = 4$ occur twice. Repeated terms in j–j coupling can be distinguished by the seniority number v, just as in the case of L–S coupling. The terms of the configuration $(7/2)^4$ that are repeated can be divided into those with $v = 2$, which are obtained from the corresponding terms of the configuration $(7/2)^2$, and terms with $v = 4$, which first appear in the configuration $(7/2)^4$.

4.2 The Connection between Molecular Terms and Nuclear Spin

4.2.1 Classification of Molecular Terms and the Total Nuclear Spin

The representation of the total wave function as a product of spatial and spin wave functions allows for the derivation of simple and clear method for finding the molecular, nuclear and electronic, spin-multiplets permitted by the Pauli exclusion principle. We start with a procedure for finding the values of the total nuclear spin allowed by the Pauli exclusion principle for molecular states with a given point symmetry.

The total nuclear spin I is a good quantum number, if we neglect the hyperfine interactions. For nuclei with the half-integer spin i the total wave function can be presented in the same form as for electrons, Eq. (4.2). The function, which is antisymmetric with respect to permutations of identical nuclei, is presented similar to the electron system case:

$$\Psi^{[1^N]} = \frac{1}{\sqrt{f_\lambda}} \sum_r \Phi_r^{[\lambda]}(x, X) \Omega_r^{[\tilde{\lambda}]}(\sigma_i), \tag{4.13}$$

where x denotes the electronic, X denotes the nuclear spatial coordinates, and σ_i denotes the nuclear spin coordinates.

For nuclei with the integer nuclear spin i, the total wave function should be symmetric with respect to permutations of identical nuclei, and the permutation symmetries of spatial and nuclear spin wave functions must be the same:

$$\Psi^{[N]} = \frac{1}{\sqrt{f_\lambda}} \sum_r \Phi_r^{[\lambda]}(x, X) \Omega_r^{[\lambda]}(\sigma_i). \tag{4.14}$$

A spin wave function for a system of identical nuclei $\Omega_r^{[\lambda]}$ transforms according to the irreducible representation $U_{2i+1}^{[\lambda]}$ of the group of unitary transformations in the $(2i+1)$-dimensional vector space of the nuclear spins. A spin Young diagram $[\lambda]$ therefore cannot have more than $2i+1$ rows. This immediately imposes limitations upon the form of the spatial Young diagram. When i is integral the spatial Young diagram cannot have more than $(2i+1)$ rows, and when i is half-integral not more than $2i+1$ columns.

For nuclei with spin $i = 1/2$ the values of the resultant spin I correspond uniquely to the form of the Young diagram which defines the permutation symmetry of the nuclear spin wave function. There is no such unique connection for nuclei with arbitrary spin i (the situation is similar to the coupling of orbital angular momenta of equivalent electrons). The possible values of the resultant spin I are determined from the decomposition

$$U_{2i+1}^{[\lambda]} \doteq \sum_I a_\lambda^{(I)} D^{(I)}, \tag{4.15}$$

the method for which is given in Appendix C, Section C.3.4.

In the Born–Oppenheimer approximation, the spatial wave function of a molecule can be written in the form of a product of an electronic wave function with a nuclear wave function:

$$\Phi(x,X) = \Phi_{\mathrm{el}}(x,X)\Phi_{\mathrm{nuc}}(X), \tag{4.16}$$

where the electronic wave function depends upon the position of nuclei X as parameters.

If it is assumed that the relative displacements of the nuclei from their equilibrium positions are small, one may take for the electronic wavefunction its value at the equilibrium configuration of the nuclei. The spatial wavefunction for a molecule has then the form of a product of wavefunctions for each type of motion:

$$\Phi(x,X) = \Phi_{\mathrm{el}}(x,X^{\mathrm{o}})\Phi_{\mathrm{vib}}(Q)\Phi_{\mathrm{rot}}(\Theta), \tag{4.17}$$

where x stands for the electronic coordinates, X^{o} are the coordinates of the equilibrium configuration of the nuclei, Q are the vibrational normal coordinates, and Θ is the set of three Euler angles which define the orientation of the molecule in space.

Equation (4.17) proposes that the nuclei execute small vibrations about equilibrium positions corresponding to an equilibrium configuration of the molecule. In this approximation the potential energy for the system of nuclei is invariant under those permutations of identical nuclei that correspond to operations of the point symmetry of the equilibrium nuclear configuration. The electronic and vibrational energy levels are classified according to the irreducible representations $\Gamma^{(\alpha)}$ of the point symmetry group of the molecule, and the wavefunctions forming bases for these representations.

The permutations of N identical nuclei are generally not all realized by the operation of the point group, and hence this group is isomorphic with a subgroup of the permutation group for identical nuclei, π_N. In order to construct basis functions for representations of the permutation group from basis functions for its point subgroup, one must proceed from the $N!/g$ equilibrium configurations of the nuclei that do not occur in the operations of the point symmetry (g here denotes the dimension of the point group). The desired basis functions are then obtained by acting of the corresponding Young operators upon basis functions for the representations $\Gamma^{(\alpha)}$.

One is now faced with the question, which representations $\Gamma^{[\lambda]}$ of the permutations group can be constructed from a basis for a representation $\Gamma^{(\alpha)}$ of its point subgroup. The answer to this is provided by the Frobenius reciprocity theorem[6]:

[6] The proof of this theorem can be found in Ref. [6].

A reducible representation Γ of a group **G** *which can be constructed from a basis for an irreducible representation $\Gamma^{(h)}$ of a subgroup $H \in G$ contains each irreducible representation $\Gamma^{(g)}$ the same number of times, as the representation $\Gamma^{(h)}$ occurs in the decomposition of $\Gamma^{(g)}$ on reducing* $\mathbf{G} \rightarrow \mathbf{H}$

Consequently, the number of independent bases for a representation $\Gamma^{[\lambda]}$, which can be constructed from a basis for a representation $\Gamma^{(\alpha)}$, is given by the coefficients in the decomposition

$$\Gamma^{[\lambda]} \doteq \sum_{\alpha} a_{\lambda}^{(\alpha)} \Gamma^{(\alpha)}. \tag{4.18}$$

The coefficients $a_{\lambda}^{(\alpha)}$ are found by the standard procedure (see Eq. A.47) with the aid of character tables for the point and permutation groups. However, it is first of all necessary to place each operation of the point group into correspondence with the appropriate permutation of the nuclei.

The connection between the permutation symmetries of the spatial and spin functions depends upon the statistics of the nuclei (see relations (4.13) and (4.14)). The existence of such a connection, together with the decompositions (4.18) and (4.15), makes it possible to associate with each type of point symmetry $\Gamma^{(\alpha)}$ a possible value of the total nuclear spin I. The described method of finding the allowed nuclear molecular multiplets was created by the author [7] in 1959 and can be represented schematically as follows:

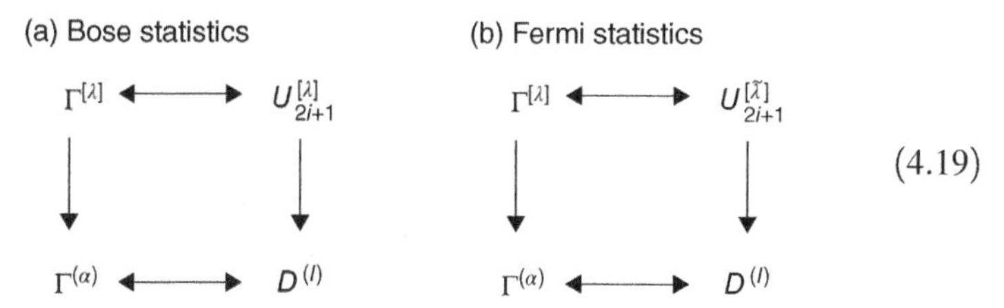

(4.19)

We now consider several examples on the application of the procedure (4.19).

EXAMPLE 1 The methane molecule $^{12}CH_4$, $i(^{12}C) = 0$, $i(H) = 1/2$; its point symmetry group is $\boldsymbol{T_d}$, it is isomorphic with the group $\boldsymbol{\pi_4}$. There is therefore a one-to-one correspondence between the classes of these groups:

$$\begin{array}{llllll} \text{Classes of } \boldsymbol{T_d}: & E & C_3 & C_2 & \sigma_d & S_4 \\ \text{Classes of } \boldsymbol{\pi_4}: & \{1^4\} & \{13\} & \{2^2\} & \{1^2 2\} & \{4\} \end{array} \tag{4.20}$$

There must also be a one-to-one correspondence between the irreducible representations of the two groups. Comparing character tables, we find

$$\begin{array}{llllll} \text{Irreducible representations of } \boldsymbol{T_d}: & A_1 & A_2 & E & F_1 & F_2 \\ \text{Irreducible representations of } \boldsymbol{\pi_4}: & [4] & [1^4] & [2^2] & [21^2] & [31] \end{array} \qquad (4.21)$$

For $i = 1/2$ and $N = 4$, the following spin Young diagrams are permitted:

$$\begin{array}{llll} [\lambda]_{\text{spin}}: & [4] & [31] & [2^2] \\ I \quad : & 2 & 1 & 0 \end{array}.$$

The spatial Young diagrams, which correspond to these, are

$$[\lambda]_{\text{spat}}: \; [1^4] \; [21^2] \; [2^2],$$

or, using Eq. (4.21), we find the corresponding irreducible representations

$$A_2, \; F_1, \; E.$$

As a result, we obtain the following allowed multiplets:

$$^5A_2, \, ^3F_1, \text{ and } ^1E, \qquad (4.22)$$

where the multiplicities $(2I+1)$ refer to the total nuclear spin.

EXAMPLE 2 In deuteromethane, $^{12}CD_4$, $i(D) = 1$, the point group is the same as that of methane, CH_4. However, the nuclei now obey the Bose statistics. We write out all the possible spin Young diagrams containing no more than three rows and the values of the total nuclear spin (these last being taken from Table C.2 in Appendix C, Section C.4).

$$\begin{array}{lllll} [\lambda]_{\text{spin}}: & [4] & [31] & [2^2] & [21^2] \\ I \quad : & 4,2,0 & 3,2,1 & 2,0 & 1 \end{array}.$$

The spatial Young diagrams coincide with those for the spins. Taking into account the one-to-one correspondence (4.21) between the irreducible representations, we obtain the following allowed multiplets:

$$^9A_1, \, ^7F_2, \, ^5A_1EF_2, \, ^3F_1F_2, \, ^1A_1E. \qquad (4.23)$$

EXAMPLE 3 The ethylene molecule, $^{12}C_2H_4$, has the point symmetry $\mathbf{D}_{2h}$. Since, as in methane, we have four fermionic nuclei H, the possible spin Young diagrams are

Table 4.3 Characters of the irreducible representations of $\boldsymbol{\pi}_4$ corresponding to operations of the group $\mathbf{D}_{2h}$

Classes of $\mathbf{D}_{2h}$	E	C_2^x	C_2^y	C_2^z	I	σ_{yz}	σ_{xz}	σ_{xy}
Classes of $\boldsymbol{\pi}_4$	$\{1^4\}$	$\{2^2\}$	$\{2^2\}$	$\{2^2\}$	$\{2^2\}$	$\{2^2\}$	$\{2^2\}$	$\{1^4\}$
$\chi^{[1^4]}$	1	1	1	1	1	1	1	1
$\chi^{[21^2]}$	3	−1	−1	−1	−1	−1	−1	3
$\chi^{[2^2]}$	2	2	2	2	2	2	2	2

$$\begin{array}{llll} [\lambda]_{\text{spin}}: & [4] & [31] & [2^2] \\ I \quad : & 2 & 1 & 0 \end{array}.$$

The dual spatial Young diagrams which correspond to these are

$$[\lambda]_{\text{spat}}: \ [1^4] \ [21^2] \ [2^2].$$

The group $\boldsymbol{\pi}_4$, however, is larger than the point group $\mathbf{D}_{2h}$, which contains eight operations. Table 4.3 gives the correspondence between the classes of these two groups and also the characters of the irreducible representations $\Gamma^{[\lambda]}$ corresponding to the operations of the point group (the C_2^x axis is chosen to be along the C=C bond, and the C_2^z axis perpendicular to the plane of the molecule). The decomposition of the representations $\Gamma^{[\lambda]}$ on the representations of the $\mathbf{D}_{2h}$ point group has the following form:

$$\Gamma^{[1^4]} \doteq A_g, \ \Gamma^{[2^2]} \doteq 2A_g, \ \Gamma^{[21^2]} \doteq B_{1g} + B_{2u} + B_{3u}. \tag{4.24}$$

One thus obtains the following nuclear multiplets for the ethylene molecule:

$$^5A_g, \ ^3B_{1g}B_{2u}B_{3u}, \ ^1A_g(2). \tag{4.25}$$

4.2.2 *The Determination of the Nuclear Statistical Weights of Spatial States*

In the previous section, it was shown how the methods of group theory allow one to determine the permitted types of spatial symmetry and the values of the total nuclear spin which correspond to them. In the zeroth approximation, when nuclear spin interactions are neglected, all the nuclear multiplets for a given symmetry type of coordinate wavefunction belong to a same energy level. The degree of degeneracy of an energy level $\Gamma^{(\alpha)}$ with respect to the nuclear spin states is called the *nuclear statistical weight* of the level, and we denote this by $\rho_{\text{nuc}}^{(\alpha)}$.

If a molecule consists of nuclei with spin $i=0$ and of a single group of identical nuclei with nonzero spin, then the nuclear statistical weight of the term $\Gamma^{(\alpha)}$ is equal to the sum of all its multiplicities, multiplied by the number of times each multiplet occurs. We denote this last quantity by $a_{2I+1}^{(\alpha)}$; that is,

$$\rho_{\text{nuc}}^{(\alpha)} = \sum_I a_{2I+1}^{(\alpha)}(2I+1). \tag{4.26}$$

This formula makes the calculation of the nuclear statistical weights of the terms easy if their multiplicities are known. As an example, we write out the nuclear statistical weights of the allowed terms for the molecules which were considered in the previous section. The value of $\rho_{\text{nuc}}^{(\alpha)}$ is given as a factor in front of the term symbol:

$$\begin{array}{lllll} {}^{12}\mathrm{CH}_4: & 5A_2, & 1E, & 3F_1; & \\ {}^{12}\mathrm{CD}_4: & 15A_1, & 6E, & 3F_1, & 15F_2; \\ {}^{12}\mathrm{C}_2\mathrm{H}_4: & 7A_g, & 3B_{1g}, & 3B_{2u}, & 3B_{3u}. \end{array} \tag{4.27}$$

If the molecule contains several distinct nuclei with spin $i_a \neq 0$ in addition to identical nuclei with spin i, the statistical weights which are obtained by taking into account the identical nuclei must be multiplied by the number of spin states of the distinct nuclei. We denote this last quantity by γ. If such nuclei are enumerated by the index a, then

$$\gamma = \prod_a (2i_a + 1). \tag{4.28}$$

The total number of spin states for a molecule is equal to

$$\rho_{\text{nuc}} = \gamma \sum_\alpha \rho_{\text{nuc}}^{(\alpha)} f_\alpha = (2i+1)^N \prod_a (2i_a + 1), \tag{4.29}$$

where $\rho_{\text{nuc}}^{(\alpha)}$ takes into account the contribution from identical nuclei only, and f_α is the dimension of the irreducible representation $\Gamma^{(\alpha)}$.

Let us consider a molecule which contains a group of identical nuclei with spin i and several individual nuclei, the number of whose spin states is equal to γ (4.28). For a system of identical nuclei one can construct $(2i+1)^N$ distinct spin functions, which can be resolved by a linear transformation into irreducible sets of functions, just as in the case of a configuration of equivalent electrons l^N (see Section 4.1.1). Each set of such functions is characterized by a particular Young diagram $[\lambda]$

consisting of N cells whose column lengths do not exceed $2i+1$, and each set contains $f_\lambda \delta_\lambda(2i+1)$ functions. These functions can be represented as points arranged in the form of a rectangular matrix:

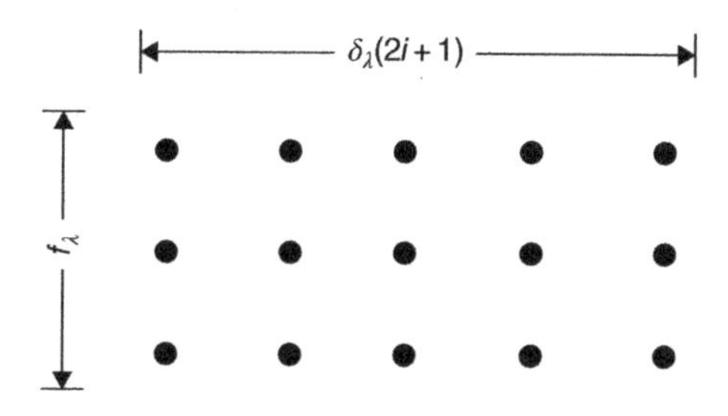

Spin functions, which lie on a single column, form a basis for an irreducible representation $\Gamma^{[\lambda]}$ of the permutations group and on constructing a total wavefunction according to formulas (4.13) and (4.14); they enter in a single linear combination. The number of distinct total wavefunctions, which can be constructed for a coordinate state $\Gamma^{[\lambda]}$, is equal to $\delta_\lambda(2i+1)$ in the case of a symmetric total wavefunction and to $\delta_{\tilde{\lambda}}(2i+1)$ in the case of an antisymmetric function.[7] It is obvious that

$$\delta_\lambda(2i+1) = \sum_I a_\lambda^{(I)}(2I+1), \tag{4.30}$$

where the $a_\lambda^{(I)}$ are determined by the decomposition (4.15). At the same time $\delta_\lambda(2i+1)$ can be calculated independently by the formula (C.76).

The number of coordinate states $\Gamma^{[\lambda]}$ which can be constructed from the coordinate states of point symmetry $\Gamma^{[\alpha]}$ is equal according to Frobenius' theorem to the coefficients $a_{\tilde{\lambda}}^{(\alpha)}$ in the decomposition (4.18). As a result we arrive at the following formulae for the nuclear statistical weights $\rho_{\text{nuc}}^{(\alpha)}$:

(a) Bose statistics (b) Fermi statistics

$$\rho_{\text{nuc}}^{(\alpha)} = \gamma \sum_\lambda a_{\tilde{\lambda}}^{(\alpha)} \delta_\lambda(2i+1) \qquad \rho_{\text{nuc}}^{(\alpha)} = \gamma \sum_\lambda a_\lambda^{(\alpha)} \delta_{\tilde{\lambda}}(2i+1) \tag{4.31}$$

For a more detailed information relating to the nuclear statistical weights, in particular, for rotational levels, see author's book [8], sections 6.7 and 6.8.

[7] We note that since all the f_λ coordinate wavefunctions, which belong to a basis for a representation $\Gamma^{[\lambda]}$, occur in a single linear combination when constructing a total wavefunction, these functions describe a single physical state.

4.3 Determination of Electronic Molecular Multiplets

4.3.1 Valence Bond Method

As is well known, the nature of the chemical bond was elucidated only after the advent of quantum mechanics and after the introduction of the concept of electron spin. However, the spins of the electrons take no direct part in the formation of molecules, the interactions that lead to chemical bonding being entirely electrostatic in nature. The dependence of the molecular energy upon the value of the total spin of the state arises from the direct connection between the permutation symmetry of the spatial wave function of the state and the value of the total spin; in Section 2.3.2 we discuss this matter in detail.

In nonrelativistic approximation the total wave function for a system of N electrons is expressed according to Eq. (4.2), as a sum of products of spatial wave functions $\Phi_{rt}^{[\lambda]}$ with spin $\Omega_{\tilde{r}}^{[\tilde{\lambda}]}$ functions, respectively, symmetrized according to mutually dual Young diagrams. The energy levels of the system of interacting electrons are characterized by the permutation symmetry $[\lambda]$ of the spatial wave function. Since every spin Young diagram $[\tilde{\lambda}]$ is uniquely associated with a value of the total spin S, one can assign to each energy level a specific value of S. All properties of particular systems that do not involve spin interactions are completely determined by the spatial wave function $\Phi_{rt}^{[\lambda]}$.

In quantum mechanical calculations one normally starts from a one-particle approximation. The one-electron orbitals can be chosen as molecular orbitals (MOs) in the form of linear combinations of atomic orbitals (*the LCAO–MO method, usually at the self-consistent field (SCF) approximation*) or orbitals, which located on the individual atoms (*the valence bond method*). The evident improvement of one-particle approximation is the construction of wave function with many different configurations of one-electron orbitals. The variation method, in which one takes into account different possible configurations of one-electron orbitals, is called a *configuration interaction* (CI) method. The detailed description of various versions of the CI method is presented in our recent review [9].

If all the configurations, which arise from the chosen set of one-electron orbitals, are taken into account, the resulting secular equation in the CI method is of very high order even for not large electron systems. However, if the antisymmetric total wave function is presented in the form (4.2), it is possible partly to diagonalize the secular equation. Since the Hamiltonian of the molecule is invariant under the permutations of electrons, the secular equation breaks up into blocks, each block corresponding to a particular irreducible representation $\Gamma^{[\lambda]}$ of the spatial wave function, or to a uniquely connected with it the total spin S. Thus, there will be a secular equation for each value of S, the dimension of which is equal to the number of ways in which the particular value of S can be realized in the system.

For symmetric molecules it is possible to diagonalize the secular equation further. For this it is necessary to construct from the wave functions corresponding to a particular spin S, that is, from the spatial functions with permutation symmetry $[\lambda]$, basis functions of irreducible representations of the point symmetry group of the molecule. We shall refer to the states of N-electron molecule of point symmetry $\Gamma^{(\alpha)}$ and total spin S as *molecular electronic multiplets* and denote them by ${}^{2S+1}\Gamma^{(\alpha)}$. The construction of the variation function so as to form eigenfunctions of the multiplets breaks up the original secular equation into secular equations, each of which corresponds to some multiplet ${}^{2S+1}\Gamma^{(\alpha)}$. The order of these secular equations is equal to the number of times a particular multiplet occurs in the decomposition of the reducible representation induced by the original set of basis functions.

It should also be mentioned that if all the configurations, which arise from the chosen set of one-electron orbitals, are taken into account, the results obtained by the MO methods and the valence bond method are equivalent, see Slater [10] and a model example of detailed calculation of H_2 in sections 8.1 and 8.2 of my book [8]. Further, we will not consider the CI approach. In this section we will discuss how to find the possible molecular multiplets in the simple version of the valence bond method [11].

Let us consider the case where there is only one nondegenerate one-electron orbital per atom, the number of valence electrons being equal to the number of orbitals. This situation occurs in calculations of the π-electron systems of conjugated and aromatics hydrocarbons and also in problems with s electrons. Each orbital is singly occupied in a configuration of neutral atoms. The interactions of the valence electrons in such a configuration lead to the formation of *covalent chemical bonds*. A configuration of singly occupied orbitals, characterized by a certain mode of coupling the spins of electrons to form the total spin S, is called a *covalent structure*. Usually there are several independent ways of coupling the electron spins to give a particular value of the total spin. The number of such ways determines the number of independent covalent structures with spin S that can be constructed from N valence electrons. We denote this number by $n(N, S)$.

In order to find the independent covalent structures, one makes use of the so-called *Rumer's rule*, see Eyring et al. [12]. According to this rule, we place symbols representing the orbitals of the valence electrons around the circumference of a circle and join pairs of orbitals by bonding lines in all possible ways without any bonds intersecting one another. A bond joining two orbitals signifies that the spins of the participating electrons are paired. For example, for four orbitals there are two independent covalent structures with $S=0$:

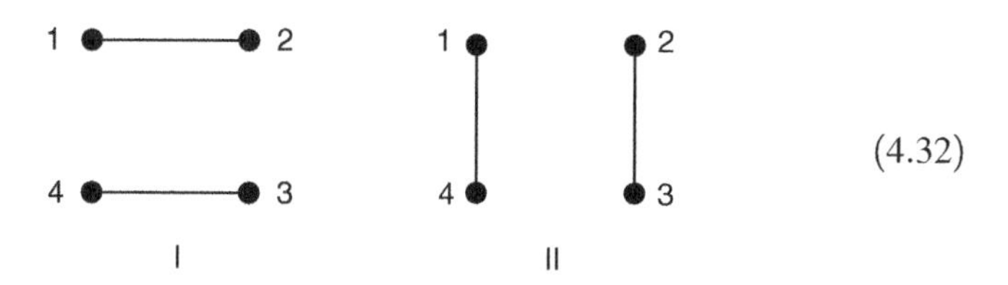

(4.32)

Usually in order to construct the wavefunction corresponding to a covalent structure, one forms the particular linear combination of determinants of spin orbitals which corresponds to the given type of coupling of electron spins [12, 13]. However, the use of the theory of the permutation group enables one to write the total wavefunction for a system of electrons in terms of products of appropriately symmetrized spatial wavefunctions with spin functions. This makes a different approach to the description of covalent structures possible, based upon the use of spatial wavefunctions. This approach is called in general as *quantum chemistry without spin*, see Chapter 1, p. 14, and the discussion at the end of Chapter 2.

In the zero approximation in interactions, the spatial wavefunction for a system of valence electrons is written as a product of N one-electron orbitals, each of which is localized on a particular atom,

$$\Phi_0 = \varphi_1(1)\varphi_2(2)\ldots\varphi_N(N). \tag{4.33}$$

In the considered case of singly occupied atomic orbitals, that is, when all φ_a in (4.33) are different, the configuration corresponds to neutral atoms. We assume that the orbitals constitute an orthonormal set. The construction of basis functions for irreducible representations $\Gamma^{[\lambda]}$ of the permutation group π_N is carried out by applying the Young operators $\omega_{rt}^{[\lambda]}$ to the function (4.33), see Section 2.3.2. As we discuss there, the f_λ^2 functions $\Phi_{rt}^{[\lambda]}$, which are obtained in this way, are divided into f_λ sets, the sets being enumerated by the index t. Each set consists of f_λ functions $\Phi_{rt}^{[\lambda]}$ with a fixed index t, and forms a basis for an irreducible representation $\Gamma^{[\lambda]}$. All functions, which belong to a same basis, correspond to one physical state, because when the total wavefunction is constructed they all occur in a single antisymmetric combination:

$$\Psi_t^{[\lambda]} = \frac{1}{\sqrt{f_\lambda}} \sum_r \Phi_{rt}^{[\lambda]} \Omega_{\tilde{r}}^{[\tilde{\lambda}]}. \tag{4.34}$$

For a given Young diagram $[\lambda]$, the different states are enumerated by the index t which characterizes the symmetry of a function (4.34) with respect to permutations of the orbitals. The number of distinct states of permutation symmetry $[\lambda]$ is therefore equal to the dimension f_λ of the irreducible representation $\Gamma^{[\lambda]}$. This is also the number of independent covalent structures with spin S, because the functions (4.34) describe a configuration of neutral atoms in a state with a particular value of the spin S.

We now express f_λ in terms of the number of valence electrons N and the value of the spin S. For this purpose we note that the dimensions of the representations $\Gamma^{[\lambda]}$ and $\Gamma^{[\tilde{\lambda}]}$ are the same. According to Eq. (2.56), the spin S is connected with rows of a spin Young diagram as

$$S = \frac{1}{2}\left(\tilde{\lambda}^{(1)} - \tilde{\lambda}^{(2)}\right). \tag{4.35}$$

And, in addition,

$$N = \tilde{\lambda}^{(1)} + \tilde{\lambda}^{(2)}. \tag{4.36}$$

Relations (4.35) and (4.36) enable one to express the row lengths of a spin Young diagram in terms of the two parameters N and S:

$$\tilde{\lambda}^{(1)} = \frac{1}{2}N + S, \quad \tilde{\lambda}^{(2)} = \frac{1}{2}N - S. \tag{4.37}$$

Substituting these values into the formula for the dimension of an irreducible representation, Eq. (B.21), we obtain the desired expression:

$$n(N,S) \equiv f_\lambda = \frac{N!(2S+1)}{((N/2)+S+1)!((N/2)-S)!}. \tag{4.38}$$

For $N=4$, $S=0$, this formula gives $n(4,0)=2$, in agreement with (4.32).

In calculations on the lower electronic states of the benzene molecule, one often takes into account just the six π electrons of the benzene ring. The number of covalent structures with $S=0$ is equal to $n(6,0)=5$. If the five independent covalent structures are written out by means of Rumer's rule, one sees that they are just the two Kekulé and three Dewar structures so well known in organic chemistry. The number of covalent structures with $S=1$ is equal to $n(6,1)=9$, the number of independent covalent structures for higher multiplicities being found similarly.

The wavefunctions for the covalent structures are linear combinations of the functions (4.34). Since the roots of a secular equation are invariant with respect to a linear transformation of the basis functions, calculations with the functions (4.34) are completely equivalent to calculations with covalent structure functions. However, the use of the functions $\Phi_{rt}^{[\lambda]}$ makes the calculation of matrix elements considerably simpler and more systematic [8].

Additional degeneracy occurs in symmetric molecules due to the symmetry of the molecular potential field in which the electrons are moving. All the $f^{(\alpha)}$ states, which belong to a single irreducible representation $\Gamma^{(\alpha)}$ of the point symmetry group of the molecule, belong to a single energy level. In order to find which irreducible representations $\Gamma^{(\alpha)}$ can occur with a particular permutation symmetry $\Gamma^{[\lambda]}$, we use the fact that a discrete point symmetry group for a molecule is isomorphic with some subgroup of the permutation group of the orbitals. As a result, any operations of the point symmetry can be represented as some permutation of the orbitals, since each orbital is localized on a particular atom. On passing from the complete permutation group to its point subgroup, a representation $\Gamma^{[\lambda]}$ breaks up into irreducible representations $\Gamma^{(\alpha)}$, which are found by decomposing the characters. Thus, with each $\Gamma^{(\alpha)}$ there is associated a value of the spin S, which corresponds uniquely to the permutation symmetry $[\tilde{\lambda}]$ of the spin wavefunction.

The method of finding the possible electronic multiplets for covalent structures is thus similar to the method of determining the nuclear molecular multiplets in case of the Fermi statistics (4.19b) and can be represented diagrammatically as follows:

$$\begin{array}{ccc} \Gamma^{[\lambda]} & \longleftrightarrow & U_2^{[\tilde{\lambda}]} \\ \updownarrow & & \updownarrow \\ \Gamma^{(\alpha)} & \longleftrightarrow & D^{(S)} \end{array} \tag{4.39}$$

However, for electrons there is a simplification because of the unique connection between $U_2{}^{[\lambda]}$ and S.

Scheme (4.39) is based upon the assumption that the particular orbitals on the atoms are of a single type, which remains invariant under the operations of point group, so that an arbitrary operation of the group is equivalent to a permutation of the orbitals. A connection is thus established between the permutation symmetry of the spatial wavefunction and its point symmetry. In the case of ionic configurations the procedure becomes a little more complicated. It is discussed in Ref. [11]; see also section 6.10 in book [8].

The method described above for neutral covalent configurations and its modification for ionic configurations was applied to the problem of finding the allowed multiplets for the ring of six hydrogen atoms. This problem was solved by Mattheiss [14] and its solution presented in detail in Slater's book [10]. The method they used was based on the distribution of electron spins among the atoms in all possible ways and finding the characters of the reducible representations, followed by subtraction from one another of the representations corresponding to different values of the total spin projections. This method is very cumbersome even for finding the 14 multiplets arising from the covalent structures. The application of the methodology described above allowed finding all 268 multiplets which arise in this problem, without any difficulties, see Ref. [11] and section 6.11 in book [8].

In the following text, we will consider more general cases of degenerate orbitals and several valence electrons per atom [15]. We will derive the explicit formulae for characters of reducible representations of the point symmetry group of the molecule induced by specified atomic states. Decomposition of these representations on irreducible parts permitted to find the Pauli-allowed molecular multiplets for arbitrary many-atom molecule.

The problem of finding the allowed molecular multiplets arising from given atomic ones was first considered by Kotani [16]. However, the general formulae given by Kotani for characters of the sought reducible representation were expressed not in terms of the characters of the irreducible representations of the group of orthogonal transformations in three-dimensional space, but in terms of the matrix elements of the corresponding matrices of this group. In addition, the application of the Kotani method required cumbersome calculations of the spin factor. The approaches developed by author and Rodimova [15, 17, 18] greatly simplify the procedure of finding the characters of representations formed by a given set of atomic states, see subsequent sections.

4.3.2 Degenerate Orbitals and One Valence Electron on Each Atom

Assume that there are N equivalent atoms, on each of which an identical degenerate orbital $\varphi_m^{(l)}$ is specified, where l is the orbital angular momentum and m is its projection on the z-axis. The number of valence electrons is assumed equal to the number of atoms. In this case the total number of orbital states is equal to $(2l+1)^N$. Each of them corresponds to a product of nonsymmetrized spatial one-electron functions,

$$\Phi_{m_1 m_2 \ldots m_N}^{(l)} = \varphi_{m_1 a}^{(l)}(1)\varphi_{m_2 b}^{(l)}(2)\ldots\varphi_{m_N q}^{(l)}(N), \tag{4.40}$$

where indices $a, b, \ldots, q$ identify the atoms. All the possible permutations of the arguments in the function (4.40) produce $N!$ functions for each set of values m_1, $m_2, \ldots, m_N$. Using the Young operators $\omega_{rt}^{[\lambda]}$, Eq. (B.33), we obtain, as many times before in this book, the basis functions of the irreducible representation $\Gamma^{[\lambda]}$ of the permutation group π_N

$$\Phi_{rt}^{[\lambda]}\left(\varphi_{m_1 a}^{(l)}\varphi_{m_2 b}^{(l)}\cdots\varphi_{m_N q}^{(l)}\right) = \omega_{rt}^{[\lambda]}\Phi_{m_1 m_2 \ldots m_N}^{(l)}. \tag{4.41}$$

For each set of values $m_1, m_2, \ldots, m_N$ there are f_λ^2 functions (4.41). These functions break up into f_λ independent basis sets, characterized by index t, with f_λ functions in each set. Under the action of permutations, the functions (4.41) with fixed t and $m_1, m_2, \ldots, m_N$ are transformed into each other. Consequently, the number of independent states described by functions (4.41) is equal to $f_\lambda(2l+1)^N$.

The action of an operation of the point symmetry group of the molecule (we denote such an operation, specified at the origin of the molecular coordinate system, by $\mathfrak{R}$) on the configuration of localized orbitals reduces to a permutation $\bar{P}$ of the centers of the orbitals and to operations of the point symmetry, specified at the center of each orbital (we denote such operations by R_a). With respect to the point group, the set of functions (4.41) with fixed λ and r forms a basis for a certain reducible representation of dimension $f_\lambda(2l+1)^N$.

The operations of the point symmetry and permutations commute. Therefore,

$$\begin{aligned}\Re\Phi_{rt}^{[\lambda]}\left(\varphi_{m_1 a}^{(l)}\varphi_{m_2 b}^{(l)}\cdots\varphi_{m_N q}^{(l)}\right) &= \Phi_{rt}^{[\lambda]}\left(\bar{P}\left(R_a\varphi_{m_1 a}^{(l)}\right)\left(R_b\varphi_{m_2 b}^{(l)}\right)\cdots\left(R_q\varphi_{m_N q}^{(l)}\right)\right) \\ &= \sum_{m_1' m_2' \ldots m_N'} D_{m_1' m_1}^{(l)}(R) D_{m_2' m_2}^{(l)}(R)\ldots D_{m_N' m_N}^{(l)}(R)\Phi_{rt}^{[\lambda]}\left(\varphi_{m_1'\bar{a}}^{(l)}\varphi_{m_2'\bar{b}}^{(l)}\cdots\varphi_{m_N'\bar{q}}^{(l)}\right)\end{aligned} \tag{4.42}$$

where $D_{m_i' m_i}^{(l)}(R)$ are the matrix elements of the irreducible representations of the group of orthogonal transformations in the three-dimensional space $\mathbf{O_3}$ corresponding to the operations R of the point group. The permutations $\bar{P}$ are defined as

$$\bar{P} = \begin{pmatrix} a\, b\cdots q \\ \bar{a}\, \bar{b}\cdots \bar{q}\end{pmatrix}.$$

Let us denote by P the permutations of electron coordinates that return the electrons to "their own" atoms. Then the function on the right-hand side of Eq. (4.42) can be represented as

$$\begin{aligned}\Phi_{rt}^{[\lambda]}\left(\varphi_{m_1'\bar{a}}^{(l)}\varphi_{m_2'\bar{b}}^{(l)}\cdots\varphi_{m_N'\bar{q}}^{(l)}\right) &= \omega_{rt}^{[\lambda]}\varphi_{m_1'\bar{a}}^{(l)}(1)\varphi_{m_2'\bar{b}}^{(l)}(2)\ldots\varphi_{m_N'\bar{q}}^{(l)}(N) \\ &= \omega_{rt}^{[\lambda]}P^{-1}\varphi_{\bar{m}_1' a}^{(l)}(1)\varphi_{\bar{m}_2' b}^{(l)}(2)\ldots\varphi_{\bar{m}_N' q}^{(l)}(N).\end{aligned} \tag{4.43}$$

Using relation (B.37) from Appendix B, one can write

$$\omega_{rt}^{[\lambda]}P^{-1} = \sum_u \Gamma_{tu}^{[\lambda]}(P^{-1})\omega_{ru}^{[\lambda]} = \sum_u \Gamma_{ut}^{[\lambda]}(P)\omega_{ru}^{[\lambda]}, \tag{4.44}$$

Substituting (4.43) and (4.44) in (4.42), we get

$$\Re\Phi_{rt}^{[\lambda]}\left(\varphi_{m_1 a}^{(l)}\varphi_{m_2 b}^{(l)}\cdots\varphi_{m_N q}^{(l)}\right) = \sum_u \sum_{\bar{m}_1' \ldots \bar{m}_N'} \Gamma_{ut}^{[\lambda]}(P) D_{\bar{m}_1' \bar{m}_1}^{(l)}(R)\ldots D_{\bar{m}_N' \bar{m}_N}^{(l)}(R)\Phi_{ru}^{[\lambda]}\left(\varphi_{\bar{m}_1' a}^{(l)}\cdots\varphi_{\bar{m}_N' q}^{(l)}\right) \tag{4.45}$$

The matrix elements $D^{(l)}_{\bar{m}_i'\bar{m}_i}$ in (4.45) are regrouped in such a way that the order of their arrangement corresponds to the order of the arrangement of $\varphi^{(l)}_{\bar{m}_1'a}\cdots\varphi^{(l)}_{\bar{m}_N'q}$, and the sum over $m_1',\ldots,m_N'$ is replaced by the equivalent summation over $\bar{m}_1',\ldots,\bar{m}_N'$.

In order to obtain the character of representation it is necessary to take the "diagonal" term in the sum (4.45), that is, to equate $u=t$, $\bar{m}_1'=m_1,\ldots,\bar{m}_N'=m_N$, and sum over t and over all the sets m_1, m_2,..., m_N. As a result, we obtain

$$X^{[\lambda]}(\mathfrak{R})=\chi^{[\lambda]}(P)\sum_{m_1\ldots m_N} D^{(l)}_{m_1\bar{m}_1}(R)\ldots D^{(l)}_{m_N\bar{m}_N}(R). \tag{4.46}$$

This expression for the character can be easily modified in a more convenient form. For this, we take into account that the permutation $\bar{P}$ corresponding to the operation $\mathfrak{R}$ can always be represented as a product of cycles, see Appendix B. Accordingly, all atoms in the molecule break up into cycles, in which the atoms go one into the other. The sequence of the values $\bar{m}_1,\bar{m}_2,\ldots,\bar{m}_N$ is made up of $m_1,m_2,\ldots,m_N$ by the action of the permutation P^{-1}, which has the same cycle structure as $\bar{P}$. Therefore, the products $D(R)$ in (4.46) can be broken up on cyclic aggregate making it possible to express them in terms of the characters. Let us examine, for instance, the part of the sum in (4.46) that corresponds to a cycle of length k (we renumber successively the indices in the cycle),

$$\sum_{m_1\ldots m_k} D^{(l)}_{m_1m_2}(R)D^{(l)}_{m_2m_3}(R)\ldots D^{(l)}_{m_km_i}(R)=\sum_{m_1} D^{(l)}_{m_1m_1}(R^k)=\chi^{(l)}(R^k). \tag{4.47}$$

Thus, this cycle can be represented as a character. For the operation $\mathfrak{R}$ corresponding to the permutation $\bar{P}$ with the cyclic structure $\{1^{\nu_1}2^{\nu_2}\ldots k^{\nu_k}\}$ the formula for the character assumes the form

$$X^{[\lambda]}(\mathfrak{R})=\chi^{[\lambda]}(P)\left[\chi^{(l)}(R)\right]^{\nu_1}\left[\chi^{(l)}(R^2)\right]^{\nu_2}\ldots\left[\chi^{(l)}(R^k)\right]^{\nu_k}. \tag{4.48}$$

In the case when different types of orbitals are specified on atoms, several covalent configurations are possible, differing in the permutations of orbitals. Since the different covalent configurations go over into each other under the operations of the point group, the only configurations contributing to the character are those that go over into themselves under the operation $\mathfrak{R}$. We denote the number of such configurations by $\tau(\mathfrak{R})$. The procedure for finding $\tau(\mathfrak{R})$ coincides with the procedure described in detail in Ref. [11] for finding the number of ionic structures remaining invariant under the action of the operation $\mathfrak{R}$. So, in the case of different orbitals on atoms, the formula for the character takes the form

$$X^{[\lambda]}(\mathfrak{R})=\chi^{[\lambda]}(P)\chi^{(l_1)}(R^{n_1})\chi^{(l_2)}(R^{n_2})\ldots\chi^{(l_k)}(R^{n_k})\tau(\mathfrak{R}). \tag{4.49}$$

Some of the orbital momenta l_i, as well as the lengths of the cycles n_i in formula (4.49), may coincide.

The characters of the operations of the point group corresponding to the irreducible representations of the group $\mathbf{O_3}$ can be found from the formulas presented in book [1], section 98; see also Appendix C. According to Eq. (C.40), the character of the operation of rotation through an angle φ is

$$\chi^{(l)}(C_\varphi) = \frac{\sin(l+1/2)\varphi}{\sin(\varphi/2)}. \tag{4.50}$$

The character of the operation of inversion, see Ref. [1], is

$$\chi^{(l)}(I) = \pm(2l+1). \tag{4.51}$$

Plus is for the even states and minus is for the odd states. The characters of reflection in the plane σ and of mirror rotation S_φ are found by representing these operations in the form $\sigma = IC_2$ and $S_\varphi = IC_{\pi+\varphi}$, which leads to the following formulas:

$$\chi^{(l)}(\sigma) = \pm\chi^{(l)}(C_2), \;\; \chi^{(l)}(S_\varphi) = \pm\chi^{(l)}(C_{\pi+\varphi}), \tag{4.52}$$

where sign depends on the parity of the state.

Let us consider as an example a symmetrical system of three equivalent atoms with p-orbitals specified on two of them and s-orbital on the third. For this system three covalent configurations are possible.

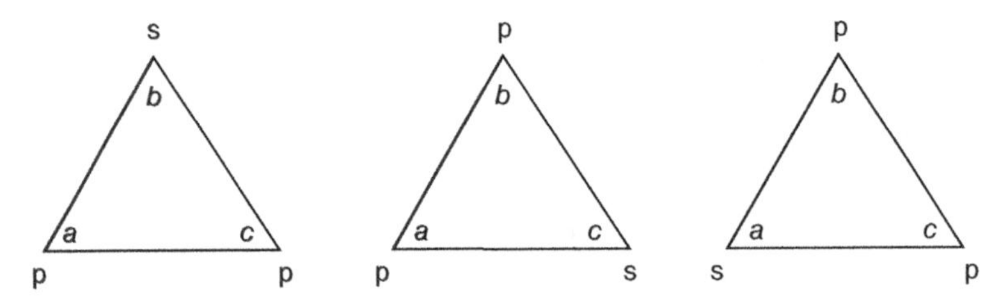

Assume that we are interested in the state of the system with the total spin $S = 3/2$. In this case the permutation symmetry of the spatial wave function is characterized by the Young diagram $[1^3]$. The point symmetry for this system is $\mathbf{C_{3v}}$. This group and the group of permutations of three triangle vertices $\boldsymbol{\pi}_3$ are isomorphic that means that between their elements and between classes of operations as well the one-to-one correspondence exists. The point symmetry group $\mathbf{C_{3v}}$ contains three classes of elements. For the class of identical operation E all three covalent configurations depicted above make the contribution to character, that is, $\tau(E) = 3$. Each of the operations of the class σ_v leaves invariant one covalent configuration, therefore, $\tau(\sigma_v) = 1$. Finally, under the action of the operations of class C_3 the configurations go over into each other, that is, $\tau(C_3) = 0$.

Table 4.4 The characters of the reducible representations and its factors for the equilateral triangle system with three identical atoms with s, p, p valence electrons on atoms in the state with $S=3/2$

Classes $\mathbf{C}_{3v}$	E	$2C_3$	$3\sigma_v$
Classes $\boldsymbol{\pi}_3$	$\{1^3\}$	$\{3\}$	$\{12\}$
$\chi^{[1^3]}$ (P)	1	1	−1
$\tau(\mathfrak{R})$	3	0	1
$\chi^{(s)}(\mathfrak{R})$	1	1	1
$\chi^{(p)}(\mathfrak{R})$	9	0	3
$X^{[1^3]}(\mathfrak{R})$	27	0	−3

The characters of representation made up by the s-orbital are always equal to unity. As for characters of representation made up by two p-orbitals their values depend on the type of operation. For the identical operation both orbitals contribute to the character, which is formed by $\chi^{(p)}(E)=3$, that is, the total contribution of the p-orbitals for the operation E is 9. For the reflection in the plane σ_v one p-orbital go over into other and because it corresponds to a cycle with the length 2, the contribution is given by $\chi^{(p)}(\sigma^2=E)=3$. The rotation C_3 gives zero contribution. The characters of the permutation group can be taken from the tables of characters presented in book [8], appendix 4. For the antisymmetric representation the characters equal the parities of permutations. This allows finding the characters of the reducible representation $X^{[1^3]}(\mathfrak{R})$. They are presented in the last line of Table 4.4 together with all characters entering in Eq. (4.49) for considered system. Decomposition of reducible representation $X^{[1^3]}(\mathfrak{R})$ on irreducible representations of the $\mathbf{C}_{3v}$ group yields the following allowed quartets: $3\,^4A_1$, $6\,^4A_2$, and $9\,^4E$.

4.3.3 Several Electrons Specified on One of the Atoms

In the preceding subsection, we considered the case when on each atom only one valence electron is specified. Formula (4.49) can be easily transformed to the case when n_1 electrons are specified on one of the atoms and all others have one valence electron. This many-electron atom is in a state characterized by quantum numbers $[\lambda_1]\alpha_1 L_1 M_1$, where the permutation symmetry of the spatial wave function, described by the Young diagram $[\lambda_1]$, is uniquely connected with the total spin of atomic state and α_1 distinguishes between states with identical values of the total orbital momentum L_1, see Section 4.1.2.

In all symmetry transformations, the many-electron atom remains in place, and only monovalent atoms are permuted. We denote such permutations by P_2. In Eq. (4.49) it is necessary to replace $\chi^{[\lambda]}(P)$ by $\chi^{[\lambda_2]}(P_2)$, where $[\lambda_2]$ denotes the permutation symmetry of the electrons of monovalent atoms, and to sum the expression for character (4.49) over all possible $[\lambda_2]$ at the fixed $[\lambda]$ and $[\lambda_1]$. As a result, Eq. (4.49) goes over into

$$X^{[\lambda]}(\mathfrak{R}) = \sum_{\lambda_2} \chi^{[\lambda_2]}(P_2)\chi^{(L_1)}(R^{n_1})\chi^{(l_2)}(R^{n_2})\ldots\chi^{(l_k)}(R^{n_k})\tau(\mathfrak{R}). \tag{4.53}$$

The direct way for finding the possible $[\lambda_2]$ is to expand the representation $\Gamma^{[\lambda]}$ of group $\boldsymbol{\pi}_N$ on its subgroup $\boldsymbol{\pi}_{n_1} \times \boldsymbol{\pi}_{n_2}$ at fixed $\Gamma^{[\lambda_1]}$:

$$\Gamma^{[\lambda]} \to \sum_{\lambda_2} \Gamma^{[\lambda_1]} \times \Gamma^{[\lambda_2]} \tag{4.54}$$

This expansion can be found by the aid of the Littlewood theorem, see Section C.3.3 in Appendix C. Alternatively, sometimes a more simple method, based on the one-to-one correspondence between Young diagram and the value of the total spin S, can be used. For finding possible $[\lambda_2]$ it is sufficient to find the possible values of the spin S_2 which satisfy the vector equations $\mathbf{S} = \mathbf{S}_1 + \mathbf{S}_2$ at fixed $\mathbf{S}_1$.

As an example, let us consider the determination of allowed triplets in the methane molecule CH_4 in the case of the carbon atom in the state $2s2p^3\ {}^3P$, that is, with $[\lambda_1] = [21^2]$, $L_1 = 1$. The electrons of the H atoms are assumed to be in the 1s state. The permutation symmetry of the total wave function is characterized by the Young diagram $[\lambda] = [2^3 1^2]$. For this $[\lambda]$ and $[\lambda_1] = [21^2]$, all possible Young diagrams for four electrons of atoms H are permitted, that is, $[\lambda_2] = [2^2]$, $[21^2]$, and $[1^4]$. In other words, the total spin $S = 1$ can be obtained by vector adding to $S_1 = 1$ all three possible values of S_2, namely, $S_2 = 0$, 1, and 2. The point symmetry group for CH_4 is $\mathbf{T}_d$. Because all H are in the same state, there is only one covalent configuration, therefore $\tau(\mathfrak{R}) = 1$ for all $\mathfrak{R}$. Taking into account that the character of representation made up by s-orbitals $\chi^{(0)}(\mathfrak{R}) = 1$ for all $\mathfrak{R}$, Eq. (4.53) takes a very simple form

$$X^{[2^3 1^2]}(\mathfrak{R}) = \sum_{\lambda_2} \chi^{[\lambda_2]}(P_2)\chi^{(1)}(R). \tag{4.55}$$

The characters $\chi^{[\lambda_2]}(P_2)$ can be taken from appendix 4 in book [8]. This is represented in Table 4.5. The characters $\chi^{(1)}(R)$ corresponding to the operations of the group $\mathbf{T}_d$ are obtained from Eqs. (4.50) to (4.52). The parity of the state of the system of k electrons with orbital angular momenta $l_1, l_2, \ldots, l_k$ is determined by the formula [1]

$$(-1)^{l_1 + l_2 + \cdots + l_k}. \tag{4.56}$$

Table 4.5 The characters of the reducible representation of CH_4 and the characters entering in Eq. (4.55) in the state with $S=1$

Classes $\mathbf{T}_d$	E	$8C_3$	$3C_2$	$6\sigma_d$	$6S_4$
Classes $\boldsymbol{\pi}_4$	$\{1^4\}$	$\{13\}$	$\{2^2\}$	$\{1^2 2\}$	$\{4\}$
$\chi^{[2^2]}\ (P_2)$	2	−1	2	0	0
$\chi^{[21^2]}\ (P_2)$	3	0	−1	−1	1
$\chi^{[1^4]}\ (P_2)$	1	1	1	−1	−1
$\chi^{(1)}\ (R)$	3	0	−1	1	1
$X^{[2^3 1^2]}(\mathfrak{R})$	18	0	−2	−2	0

Consequently, the carbon atom with the considered electron configuration $2s2p^3$ is in the odd state. In Table 4.5 we represent the calculated characters of the reducible representation $X^{[2^3 1^2]}(\mathfrak{R})$ induced by the total wave function together with characters, which are necessary for calculation by Eq. (4.55). Decomposition of this representation on irreducible representations of the $\mathbf{T}_d$ group yields the following allowed triplets for the CH_4 molecule: $3A_2$, ^3E, 3^3F_1, and 2^3F_2. They are in full accordance with the set found in Kotani's paper [16].

The case of ionic configurations is discussed in Ref. [15]. Using Eq. (4.53) and appropriate equations for ionic configurations all 100 multiplets that can be constructed on four 2p orbitals of the O atom and two 1s orbitals of the H atoms are found for the H_2O molecule.

4.3.4 Diatomic Molecule with Identical Atoms

Let us consider a diatomic molecule consisting of identical many-electron atoms. The states of the atoms are specified, as in the preceding subsection, by the set of quantum numbers $[\lambda_i]\alpha_i L_i M_i$. The spatial wave function for each atom contains n_1 electrons and characterized by the permutation symmetry of the Young diagram $[\lambda_i]$, which corresponds to a definite value of the total spin S_i of the atom. The spatial wave function of the molecule must have the permutation symmetry of the Young diagram $[\lambda]$ with $2n_1$ boxes corresponding to the total spin S of the molecule. This molecular wave function can be constructed from the spatial atomic wave functions corresponding to the given atomic quantum states similar to that performed for the configuration of two shells $j^{n_1}j^{n_2}$ in the central field; see eq. (7.56) in book [8]:

$$\begin{aligned}&\Phi_r^{[\lambda]}(A_1|\lambda_1|\alpha_1 L_1 M_1, A_2|\lambda_2|\alpha_2 L_2 M_2)\\&= C\sum_{r_1 r_2}\sum_{Q}\langle r|Q|r_1 r_2\rangle^{[\lambda]} Q\Phi_{r_1}^{[\lambda_1]}(A_1\alpha_1 L_1 M_1)\Phi_{r_2}^{[\lambda_2]}(A_2\alpha_2 L_2 M_2),\end{aligned} \tag{4.57}$$

where C is a normalization factor and Q are the permutations of electrons between atoms A_1 and A_2. Any set of $(2n_1)!/n_1!\ n_1!$ permutations can be selected as the Q. We shall choose permutations that preserve the ascending order of numbers of electrons within each atom. The right-hand side of the matrix element $\langle r|Q|r_1r_2\rangle$ belongs to so-called *nonstandard representations* of the permutation group, see chapter 2 in book [8]. They are obtained from the standard Young–Yamanouchi representations by the transformation matrices of the permutation group introduced by the author in Ref. [19].

It is useful to recall that the spectroscopic notations of the quantum states of two-atom molecules are different from the notations used for irreducible representations of its point symmetry groups. Diatomic molecules have axial symmetry about an axis passing through the two nuclei. Therefore, the electronic terms can be classified according to the values of the projection of the total angular momentum L on this axis, which is denoted by the letter Λ. The terms with different values of $\Lambda=0$, 1, 2, 3, 4... are denoted by the capital Greek letters Σ, Π, Δ, Φ, Γ, ..., respectively. In the nonrelativistic approach, the electronic states of the molecule are also characterized by the total spin S. Thus, the electronic states of diatomic molecule form the multiplets ${}^{2S+1}\Lambda$. In the case of two identical atoms in the same states, an additional symmetry appears; terms are classified according to their parity: molecular states can be even (g) or odd (u).[8] The Σ terms are characterized also by their sign in respect to reflection in a vertical plane (e.g., one passing through the molecular axis). It can be constructed for two functions with $\Lambda=0$: one, Σ^+, is unchanged upon reflection in a vertical plane, while the other, Σ^-, changes its sign.

The homoatomic dimer has the point symmetry group $\mathbf{D}_{\infty h}$. The characters of its irreducible representations and their connection with spectroscopic notations (placed in parantheses) are represented in Table 4.6.

The characters of the reducible representation induced on the functions (4.57) were obtained in Ref. [15]. We represent them here for two cases.

1. Both identical atoms are in *the same state*. The expression for the character is equal to

$$X^{[\lambda]}(P_{A_1A_2}R_{A_1}R_{A_2}) = \langle\lambda_1\lambda_1\|P\|\lambda_1\lambda_1\rangle^{[\lambda]}\chi^{(L_1)}(R^2). \qquad (4.58)$$

   The permutation factor in Eq. (4.58) is defined on the nonstandard representations of the permutation group [19] and does not depend on the Young tableaux. It is diagonal in $[\lambda_i]$, that is, the multiplicities of both atoms must always be the same. The method of finding such permutation factors is developed in Ref. [15]; the values of the permutation factors for all possible symmetries of systems of electrons with $N \leq 8$ are tabulated (see Table 4.7).

[8] The notations g and u stem from German: *gerade* and *ungerade*, respectively.

Table 4.6 Characters of the irreducible representations of the group $D_{\infty h}$

$D_{\infty h}$	E	$2C_\varphi$	U_2	I	$2IC_\varphi$	IU_2
$A_{1g}\left(\Sigma_g^+\right)$	1	1	1	1	1	1
$A_{1u}\left(\Sigma_u^+\right)$	1	1	1	−1	−1	−1
$A_{2g}\{R_z\}\left(\Sigma_g^-\right)$	1	1	−1	1	1	−1
$A_{2u}\{z\}\left(\Sigma_u^-\right)$	1	1	−1	−1	−1	1
$E_{1g}\{R_x, R_y\}\ (\Pi_g)$	2	$2\cos\varphi$	0	2	$2\cos\varphi$	0
$E_{1u}\{x, y\}\ (\Pi_u)$	2	$2\cos\varphi$	0	−2	$-2\cos\varphi$	0
$E_{2g}\ (\Delta_g)$	2	$2\cos 2\varphi$	0	2	$2\cos 2\varphi$	0
$E_{2u}\ (\Delta_u)$	2	$2\cos 2\varphi$	0	−2	$-2\cos 2\varphi$	0
$E_{3g}(\Phi_g)$	2	$2\cos 3\varphi$	0	2	$2\cos 3\varphi$	0
$E_{3u}(\Phi_u)$	2	$2\cos 3\varphi$	0	−2	$-2\cos 3\varphi$	0
$E_{4g}(\Gamma_g)$	2	$2\cos 4\varphi$	0	2	$2\cos 4\varphi$	0
$E_{4u}(\Gamma_u)$	2	$2\cos 4\varphi$	0	−2	$-2\cos 4\varphi$	0

2. The identical atoms are in *different states* and have different sets $[\lambda_i]\alpha_i L_i$. In this case, the character for the operation $\mathfrak{R}$ containing the permutation of atoms is equal to zero.

$$X^{[\lambda]}\left(P_{A_1A_2}R_{A_1}R_{A_2}\right)=0. \tag{4.59}$$

For the operations $\mathfrak{R}$, which are not connected with the permutations of atoms, the corresponding expression for the character is

$$X^{[\lambda]}\left(R_{A_1}R_{A_2}\right)=\chi^{(L_1)}(R)\chi^{(L_2)}(R)\tau_{L_1L_2}, \tag{4.60}$$

where $\tau_{L_1L_2}=2$ for $L_1 \neq L_2$, since two configurations, differing in permutations of the orbital angular momenta, are possible; obviously that for equal angular momenta $\tau_{L_1L_1}=1$.

As an example, let us consider the N_2 molecule in the state with $S=0$ for two cases.

Case 1 Both atoms N are in the same state, namely, in $p^3\ {}^2P$ with the atomic spin 1/2, the character is given by Eq. (4.58) with the permutation factor $\langle[21][21]\|P\|[21][21]\rangle^{[2^3]}$. According to Table 4.7, this permutation factor is equal to unity. Equations (4.58) and (4.60) take on the simple form

$$X^{[2^3]}(P_{A_1A_2}R_{A_1}R_{A_2})=\chi^{(1)}(R^2), \tag{4.61}$$

$$X^{[2^3]}(R_{A_1}R_{A_2})=\left[\chi^{(1)}(R)\right]^2. \tag{4.62}$$

Table 4.7 Permutation factors for all permitted symmetries of electron systems with $N \leq 8$

		$[\lambda]$											
	$[\lambda_1]$	$N=4$			$N=6$				$N=8$				
		$[2^2]$	$[21^2]$	$[1^4]$	$[2^3]$	$[2^21^2]$	$[21^4]$	$[1^6]$	$[2^4]$	$[2^31^2]$	$[2^21^4]$	$[21^6]$	$[1^8]$
	[2]	1											
	$[1^2]$	1	−1	1									
w_2	[21]				1	−1							
	$[1^3]$				1	−1	1	−1					
	$[2^2]$								1				
	$[21^2]$								1	−1	1		
	$[1^4]$								1	−1	1	−1	1
w_3	[2]				1								
	$[1^2]$				1	0	−1	1					
w_4	[2]								1				
	$[1^2]$								1	0	0	−1	1

The characters for the operations of the molecular point group $\mathbf{D}_{\infty\mathbf{h}}$ can be found using Eqs. (4.50)–(4.52). For $L = 1$ we obtain

$$\begin{array}{ccccccc} \mathrm{D}_{\infty \mathrm{h}}: & E & C_\varphi & \sigma_v & I & IC_\varphi & I\sigma_v \\ \chi^{[2^3]}[\mathfrak{R}] & 9 & \dfrac{\sin^2(3/2)\varphi}{\sin^2(\varphi/2)} & 1 & 3 & \dfrac{\sin 3\varphi}{\sin\varphi} & 3 \end{array}. \tag{4.63}$$

For the decomposition of this representation into the irreducible representations of the group $\mathbf{D}_{\infty\mathbf{h}}$ we must take into account that $\mathbf{D}_{\infty\mathbf{h}}$ belongs to continuous groups, although it contains discrete operations also. In this case in the decomposition

$$\Gamma \doteq \sum_\beta a^{(\beta)}\Gamma^{(\beta)}, \tag{4.64}$$

the expression for the coefficient $a^{(\beta)}$ includes the integration over the continuous region of the group parameter φ and a sum over discrete operations

$$a^{(\alpha)} = \frac{1}{8\pi}\sum_R \int_0^{2\pi} \chi^{(\Gamma)}(R)\chi^{(\alpha)}(R)^* d\varphi, \tag{4.65}$$

where the sum does not include the operations E and I because they are taken into account in the integral over φ. Thus, it is summed only over four classes of $\mathbf{D}_{\infty\mathbf{h}}$, see Table 4.6. Using Eq. (4.65), in which the characters of reducible representation $\chi^{(\Gamma)}(R)$ are represented above as $X^{[2^3]}(\mathfrak{R})$, see (4.63), and the irreducible representations of $\mathbf{D}_{\infty\mathbf{h}}$ are given in Table 4.6, we find the following allowed singlet states:

$$2^1\Sigma_g^+, {}^1\Sigma_u^-, {}^1\Pi_g, {}^1\Pi_u, {}^1\Delta_g. \tag{4.66}$$

Case 2 Atoms N are in different states: $\mathrm{N}\left(p^{3\,2}P\right) + \mathrm{N}\left(p^{3\,2}D\right)$. The decomposition of the reducible representation with $[\lambda] = \left[2^3\right]$ and $\tau = 2$ with the characters found by Eqs. (4.59) and (4.60) gives the following allowed singlets:

$${}^1\Sigma_g^+, {}^1\Sigma_u^+, 2^1\Sigma_g^-, 2^1\Sigma_u^-, 3^1\Pi_g, 3^1\Pi_u, 2^1\Delta_g, 2\,{}^1\Delta_u, {}^1\Phi_g, {}^1\Phi_u. \tag{4.67}$$

* *
*

It should be mentioned that the first study of classification of the Pauli-allowed dimer states that can arise from given atomic states was performed in 1928 by Wigner and Witmer [20], see also the comprehensive review by Mulliken [21] published at that time. The results obtained by Wigner and Witmer were described in detail by Herzberg in his book [22], which has been widely used by molecular spectroscopists, and systemized by Landau and Lifshitz in their textbook [1]. It should be noted that in all these publications the tables of allowed dimer states that can be constructed from given atomic states and useful rules were formulated, but the explicit formulae for the characters similar to Eqs. (4.58)–(4.60), to the best of our knowledge, were not represented.

Below we give the rules for finding the possible allowed dimer states formulated in the Landau and Lifshitz book [1], section 80. First, we consider the case of two identical atoms in the same state and designate by N_g (N_u) the number of even (odd) terms with given values of Λ and S. In this case, the following rules connecting the parity Λ and S are valid:

$$\begin{aligned} &\Lambda \text{ is odd, then } N_g = N_u; \\ &\Lambda \text{ is even and } S \text{ is even, then } N_g = N_u + 1; \\ &\Lambda \text{ is even and } S \text{ is odd, then } N_u = N_g + 1. \end{aligned} \tag{4.68}$$

For the Σ terms the rules for the parity depend on the value of S and are different for Σ^+ and Σ^- states. Namely:

$$\begin{aligned} &\text{The parity of } \textstyle\sum^+ \text{ states is } (-1)^S; \\ &\text{The parity of } \textstyle\sum^- \text{ states is } (-1)^{S+1}. \end{aligned} \tag{4.69}$$

Thus, the terms ${}^{1,5}\Sigma_g^+$, ${}^{1,5}\Sigma_u^-$, ${}^{3,7}\Sigma_u^+$, and ${}^{3,7}\Sigma_g^-$ are permitted, while the terms ${}^{1,5}\Sigma_g^-$, ${}^{1,5}\Sigma_u^+$, ${}^{3,7}\Sigma_g^+$, and ${}^{3,7}\Sigma_u^-$ are forbidden.

If the identical atoms are in different states, the number of possible states is doubled in comparison with the number of states when the atoms are not identical. The interchange of the states of identical atoms does not change the energy. Symmetrizing and antisymmetrizing the wave function of the molecule with respect to this interchange, we obtain even and odd terms with the same Λ and S. Thus for each molecular multiplet both parities are permitted and equal numbers of even and odd multiplets ${}^{2S+1}\Lambda_{g,u}$ are realized.

It is easy to check that the allowed multiplets for N_2 found in the examples above, see (4.66) and (4.67), are satisfied by these rules. Let us stress that for atoms in different states the presented formulae for characters really give the equal numbers of even and odd multiplets.

4.3.5 *General Case I*

Let us consider a molecule consisting of an arbitrary number of atoms. The state of each atom is specified by a set of quantum numbers $[\lambda_i]\alpha_i L_i M_i$ and the total permutation symmetry of the spatial wave function is characterized by the Young diagram $[\lambda]$. For the sake of clarity we consider first a molecule consisting of three identical atoms. The result of action of the cyclic permutation $P_{A_1A_2A_3}$ on the spatial function of this three-atom molecule is

$$\begin{aligned} &P_{A_1A_2A_3}\Phi_r^{[\lambda]}((A_1[\lambda_1]\alpha_1L_1M_1,\ A_2[\lambda_2]\alpha_2L_2M_2)[\lambda_{12}]A_3[\lambda_3]\alpha_3L_3M_3) \\ &= C\sum_{r_1r_2r_3}\sum_{Q}\langle r|Q|(r_1r_2)\lambda_{12}r_3\rangle^{[\lambda]}QP^{-1}\Phi_{r_1}^{[\lambda_1]}(A_2\alpha_1L_1M_1) \\ &\quad\times\Phi_{r_2}^{[\lambda_2]}(A_3\alpha_2L_2M_1)\Phi_{r_3}^{[\lambda_3]}(A_1\alpha_3L_3M_3), \end{aligned} \tag{4.70}$$

where Q, as in Eq. (4.57), are the permutations of electrons between atoms A_1, A_2, and A_3, preserving the ascending order of numbers of electrons within each atom, and the permutation P returns the electrons to "their own" atoms, while conserving the increasing order of numbers of electrons of atoms. The right-hand side of the matrix element of Q is defined on the nonstandard representation of the permutation group π_{3n_1}, which is reduced in respect to subgroup $\pi_{n_1}\times\pi_{n_1}\times\pi_{n_1}$, and $[\lambda_{12}]$ is the intermediate Young diagram characterizing the permutation symmetry of the first two groups of electrons. The expressions for characters for a three-atom molecule and in the general case were derived in Ref. [15]. For the operation $\mathfrak{R}=P_{A_1A_2A_3}R_{A_1}R_{A_2}R_{A_3}$ the following expression for the character is valid:

$$X^{[\lambda]}(\mathfrak{R})=\sum_{\lambda_{12}}\langle(\lambda_1\lambda_1)\lambda_{12}\lambda_1\|P\|(\lambda_1\lambda_1)\lambda_{12}\lambda_1\rangle^{[\lambda]}\chi^{(L)}(R^3), \tag{4.71}$$

where permutation P can be denoted as $P_{(123)}$, which means that the electrons of the atom 1 are replaced by the electrons of the atom 2, and so on, according to the action of the cyclic permutation. The permutation factor does not depend upon the Young tableaux and are diagonal in $[\lambda_i]$. The latter means that even when the atoms are in different quantum states, the multiplicities (spins) of atoms in a cycle must be the same.

It is obvious that in the case of a cycle of n atoms we have

$$X^{[\lambda]}(\mathfrak{R})=\sum_{\lambda_{\mathrm{int}}}\langle(\lambda_1\ldots\lambda_1)\lambda_{\mathrm{int}}\|P\|(\lambda_1\ldots\lambda_1)\lambda_{\mathrm{int}}\rangle^{[\lambda]}\chi^{(L)}(R^n), \tag{4.72}$$

where λ_{int} denotes the set of $(n-2)$ intermediate Young diagrams, which are necessary for the complete description of the permutation symmetry of the state of n particles. It is convenient to designate the sum over λ_{int} in Eq. (4.72) by a special symbol

$$\sum_{\lambda_{\mathrm{int}}}\left\langle\underbrace{(\lambda_1\ldots\lambda_1)}_{n}\lambda_{\mathrm{int}}\|P\|(\lambda_1\ldots\lambda_1)\lambda_{\mathrm{int}}\right\rangle^{[\lambda]}=w_n(\lambda_1,\lambda), \tag{4.73}$$

then the expression (4.72) assumes a very compact form

$$\chi^{[\lambda]}(\mathfrak{R}) = w_n(\lambda_1, \lambda)\chi^{(L)}(R^n). \quad (4.74)$$

In the general case, the operation $\mathfrak{R}$ corresponds to a permutation of atoms with an arbitrary cyclic structure. The expression for the character, taking into account the notation (4.73), can be written in the form [15]

$$\chi^{[\lambda]}(\mathfrak{R}) = \left\{ \sum_{\lambda^{(1)} \ldots \lambda^{(k)}} \sum_{\lambda_{\text{int}}} w_{n_1}\left(\lambda_1, \lambda^{(1)}\right)\chi^{(L_1)}(R^{n_1}) \ldots w_{n_k}\left(\lambda_k, \lambda^{(k)}\right)\chi^{(L_k)}(R^{n_k}) \right\} \tau(\mathfrak{R}), \quad (4.75)$$

where n_i is the number of atoms in a cycle, in which each atom is characterized by the permutation symmetry $[\lambda_i]$ and the total orbital angular momentum L_i, and $[\lambda^{(i)}]$ is the permutation symmetry of n_i atoms that constitute the i-th cycle. Since for the given permutation symmetry $[\lambda]$ only the permutation symmetry $[\lambda_i]$ of the atoms is specified, the character (4.75) is summed over all possible permutation symmetries $[\lambda^{(i)}]$ of the cycles and also over the intermediate Young diagrams $[\lambda_{\text{int}}]$ that appeared in the construction $[\lambda]$ from $[\lambda^{(i)}]$. The possible $[\lambda^{(i)}]$ at a fixed $[\lambda]$ can be obtained by using the Littlewood theorem [23], see Section C.3.3 in Appendix C. Finally, $\tau(\mathfrak{R})$ denotes, as in Eqs. (4.49) and (4.53), the number of electronic configurations of the molecule, which remains invariant under the action of the operation $\mathfrak{R}$. For particular cases, the general expression (4.75) goes over into Eqs. (4.49), (4.53), and (4.58) of the preceding sections.

In Eq. (4.75) the permutations factors $w_{n_i}\left(\lambda_i, \lambda^{(i)}\right)$ for each cycle are calculated separately. They can be calculated directly with the aid of the transformation matrices of the nonstandard representations of the permutation group, see Ref. [8], chapter 2. In Ref. [15] the system of equations, which relates the permutation factors $w_{n_i}\left(\lambda_i, \lambda^{(i)}\right)$ with the characters $\chi^{[\lambda_i]}$ and $\chi^{[\lambda^{(i)}]}$, is formulated and permutation factors for all possible symmetries of systems of electrons with $N \leq 8$ are found, see Table 4.7.

4.3.6 *General Case II*

When the number of permuted electrons in a cycle is large, the method, described in the previous subsection, becomes quite cumbersome, because it requires the knowledge of the characters of the permutation group, the determination of which at large values of N is quite difficult. Let us recall that the method of finding the allowed nuclear molecular multiplets (described in Section 4.2) does not depend on the number of nucleons in nuclei, but only on their spin. The behavior of the total wave function in respect to the permutation of nuclei depends on the statistics of nuclei. The nuclei with an even number of nucleons have an integer value of the total spin and are bosons; the nuclei with an odd number of nucleons have a half-integer value of the total spin and are fermions. Thus, it is enough to know the value of the total spin of the permuted nuclei, just as it follows from the Pauli exclusion principle for composite particles. This idea was realized for many-electron subsystems in our next after Ref. [15] publication [17]. However, in the case of many-electron subsystems the scheme (4.19) is not valid. It becomes more complicated, since in contrary to the invariance of nuclei in respect to the operations of the molecular point symmetry group, many-electron subsystems are not invariant in respect to these operations.

We consider a system with some point-group symmetry **G**, consisting of n subsystems containing n_a particles each. The state of each subsystem is characterized by an irreducible representation $\Gamma^{(\alpha_a)}$ of the local point symmetry group $\mathbf{G_a}$ and a total spin S_α. We assume the subsystems to be identical if all the S_α coincide, while $\Gamma^{(\alpha_a)}$ can be different.

With respect to the operations of the group **G**, the many-particle subsystem can be considered as a single particle with spin S_α. Depending on whether S_α is an integer or a half-integer, the subsystems obey the Bose–Einstein or Fermi–Dirac statistics, respectively, and the total wavefunction of the system can be represented as

$$\Psi = \frac{1}{\sqrt{f_\lambda}} \sum_r \Phi_r^{[\lambda]} \Omega_{\bar{r}}^{[\bar{\lambda}]}, \tag{4.76}$$

where, as usual, $\Phi_r^{[\lambda]}$ is the spatial wave function of the system transforming as the r-th column of the representation $\Gamma^{[\lambda]}$ of the permutation group $\boldsymbol{\pi}_\mathbf{n}$ of n subsystems, $\Omega_{\bar{r}}^{[\bar{\lambda}]}$ is the spin function of the system transforming as the $\bar{r}$th column of the representation $\Gamma^{[\bar{\lambda}]}$, and f_λ is the dimension of the representation $\Gamma^{[\lambda]}$. The Young diagrams $[\bar{\lambda}]$ are equal to

$$[\bar{\lambda}] = \begin{cases} [\lambda] & \text{for boson subsystems} \\ [\tilde{\lambda}] & \text{for fermion subsystems} \end{cases} \tag{4.77}$$

where $[\tilde{\lambda}]$ is the Young diagram dual to $[\lambda]$.

As we discussed above, each subsystem in respect to permutations can be considered as a single particle. Let us denote its spatial wave function $\varphi_{i_a}^{(\alpha_a)}$ as we did it for orbitals, but instead of the value of an angular momentum l, their symmetry is characterized by the irreducible representation $\Gamma^{(\alpha_a)}$ of the local point symmetry group $\mathbf{G_a}$. For the subsystem a there are f_{α_a} spatial functions $\varphi_{i_a}^{(\alpha_a)}$ where f_{α_a} is the dimension of the representation $\Gamma^{(\alpha_a)}$.

In the zero approximation in the interaction, the spatial wave function of the system can be constructed from the simple products of the spatial functions of the subsystems,

$$\Phi_0 = \varphi_{i_1}^{(\alpha_1)} \varphi_{i_2}^{(\alpha_2)} \ldots \varphi_{i_n}^{(\alpha_n)}. \tag{4.78}$$

Altogether, it is possible to make up $f_{\alpha_1} f_{\alpha_2} \ldots f_{\alpha_n}$ products (4.78). For each of them the f_λ^2 spatial wave functions

$$\Phi_{rt}^{[\lambda]} = \omega_{rt}^{[\lambda]} \Phi_0 \tag{4.79}$$

with the permutation symmetry $[\lambda]$ can be built. The functions (4.79) with fixed index t are transformed into each other upon permutations of the subsystems and in the construction of the total wave function they enter in a single bilinear combination (4.76), that is, correspond to one physical state. Thus the total number of independent states having the permutation symmetry $[\lambda]$ is $f_\lambda f_{\alpha_1} f_{\alpha_2} \ldots f_{\alpha_n}$. The spatial functions (4.79) describing these states form a basis of a certain representation $U_{\alpha_1 \ldots \alpha_n}^{[\lambda]}$ in general case reducible in respect to the point-symmetry group $\mathbf{G}$ of the system. The decomposition of this representation gives the allowed spatial states $\Gamma^{(\alpha)}$ of the system consistent with the permutation symmetry $[\lambda]$

$$U_{\alpha_1 \ldots \alpha_n}^{[\lambda]} \doteq \sum_{\alpha} c_\alpha \Gamma^{(\alpha)}. \tag{4.80}$$

As we discussed in the previous sections, the operations $\mathfrak{R}$ of the group $\mathbf{G}$ on the functions (4.79) is equivalent to a permutation P of the subsystems and point transformations R of the subsystem wave functions. The characters of the representations induced on the functions (4.79) can be represented in the form of two factors: the permutation factor and the factor depending on the point symmetry transformations (let us call it as the orbital factor). Namely:

$$X^{[\lambda]}(\mathfrak{R}) = X^{[\lambda]}(P) X^{(\alpha_1 \ldots \alpha_n)}(\mathfrak{R}). \tag{4.81}$$

The permutation factor in Eq. (4.81) is equal to the character $\chi^{[\lambda]}$ of the representation $\Gamma^{[\lambda]}$ of the permutation group of the subsystems $\boldsymbol{\pi_n}$; in contrary to the method described in preceding subsection, it does not depend upon the number of particles in subsystems and depends only on the number of subsystems,

$$X^{[\lambda]}(P) = \chi^{[\lambda]}(P). \tag{4.82}$$

The form of the orbital factor is determined by the cyclic structure of the permutation P. For P with the cyclic structure $\{1^{n_1}2^{n_2}\ldots k^{n_k}\}$, the orbital factor can be represented similar to that in Eq. (4.48) as

$$X^{(\alpha_1\ldots\alpha_n)}(\mathfrak{R}) = \left[\chi^{(\alpha_1)}(R)\right]^{n_1}\left[\chi^{(\alpha_2)}(R^2)\right]^{n_2}\ldots\left[\chi^{(\alpha_k)}(R^k)\right]^{n_k}\tau(\mathfrak{R}), \tag{4.83}$$

where $\chi^{(\alpha_i)}(R)$ is the character of the representation $\Gamma^{(\alpha_i)}$ of the local symmetry group of the subsystems contained in the cycle $\{i\}$. The representations $\Gamma^{(\alpha_i)}$ of all subsystems that enter in the cycle $\{i\}$ must coincide, otherwise the character vanishes; $\tau(\mathfrak{R})$ is the number of conformations that are invariant under the operation $\mathfrak{R}$. If $\Gamma^{(\alpha_i)}$ coincide for all subsystems, then $\tau(\mathfrak{R}) = 1$.

Our system consists of n subsystems, each of which we consider as single particles with spin S_α. The spin wave function of the system belongs to the irreducible representation $U_{2S_\alpha+1}^{[\lambda]}$ of the unitary transformation group $\mathbf{U}_{2S_\alpha+1}$. The group of three-dimensional rotations $\mathbf{R}_3$ is a subgroup of $\mathbf{U}_{2S_\alpha+1}$. So, while limiting unitary transformations by rotations in three-dimensional space, the representations $U_{2S_\alpha+1}^{[\lambda]}$ become in the general case reducible and decompose into irreducible representations $D^{(S)}$ of the group $\mathbf{R}_3$:

$$U_{2S_\alpha+1}^{[\tilde{\lambda}]} = \sum_S \alpha\left(U_{2S_\alpha+1}^{[\tilde{\lambda}]} \rightarrow D^{(S)}\right)D^{(S)}, \tag{4.84}$$

where the coefficient α is equal to the number of occurrence of representation $D^{(S)}$ in the expansion (4.84). The α-coefficients and allowed values of S for different $[\lambda]$ and S_α are represented in Tables of the reduction of the representations $U_{2j+1}^{[\lambda]}$ to the group $\mathbf{R}_3$, see Appendix C, Section C.4.

Thus, the procedure for finding the allowed multiplets of the system can be represented schematically in the following form:

$$\begin{array}{ccc} U_{\alpha_1\ldots\alpha_n}^{[\lambda]} & \longleftrightarrow & U_{2S_\alpha+1}^{[\tilde{\lambda}]} \\ \downarrow & & \downarrow \\ \Gamma^{(\alpha)} & \longleftrightarrow & D^{(S)} \end{array} \tag{4.85}$$

The expression for the character of the representation containing all terms with the given S can be written, as in (4.81), as a product of two factors:

$$X^{(S)}(\mathfrak{R}) = X^{(S)}(P)X^{(\alpha_1\ldots\alpha_n)}(\mathfrak{R}), \tag{4.86}$$

where for the orbital factor $X^{(\alpha_1 \ldots \alpha_n)}(\mathfrak{R})$, corresponding to permutation P, the expression (4.83) is valid, while the permutation factor is equal to

$$X^{(S)}(P) = \sum_{\lambda} \chi^{[\lambda]}(P)\, \alpha\left(U^{[\lambda]}_{2S_a+1} \to D^{(S)}\right), \tag{4.87}$$

where the sum is taken over all permitted $[\lambda]$ with n boxes, in reality over $[\lambda]$, which contain in the reduction (4.84) the given S. For practical convenience the permutation factor (4.87) for cycles with the number of subsystems $n = 2-4$ and spins $S_a = 1/2\ (1/2)\ 5/2$ are tabulated in Ref. [18]. Alternative methods for calculating the permutation factor, including the operation of plethysm, introduced by Littlewood [23], or an application of $3nj$-symbols, are discussed in detail in Ref. [18].

The classification described above in this and previous sections was developed for nonrelativistic systems when spin is a good quantum number and the total wave function can be represented as a product of spatial and spin wave functions of the system (Eq. 4.76). When the spin–orbit interaction is large, the states of the system are characterized only by the total wave function, which can be either symmetric or antisymmetric, depending on whether the spin S_a is an integer or a half-integer. The scheme (4.85) must be replaced by its left-hand side, but with the total wave function, for which $[\lambda] = [n]$ or $[1^n]$

$$U^{[n],[1^n]}_{\alpha_1 \ldots \alpha_n} \to \Gamma^{(\alpha)}. \tag{4.88}$$

The reduction of the representation of the unitary group in the scheme (4.88) on the irreducible representations $\Gamma^{(\alpha)}$ of the point symmetry group $\mathbf{G}$ of the system is performed according to Eq. (4.80).

The cases of impurity ion complexes in crystal field for strong and weak crystal fields are considered in Ref. [24]. The expressions for characters for systems consisting of several groups of identical subsystems are obtained in Refs. [17, 18]; there it can also be found different examples on the application of the group-theoretical classification method described in this section.

References

[1] L.D. Landau and E.M. Lifshitz, *Quantum Mechanics, (Nonrelativistic Theory)*, 3rd edn., Pergamon Press, Oxford, 1977.
[2] A. de-Shalit and I. Talmi, *Nuclear Shell Theory*, Academic Press, New York, 1963.
[3] G. Racah, *Phys. Rev.* **75**, 1352 (1949).
[4] B.R. Judd, *Operator Techniques in Atomic Spectroscopy*, McGraw-Hill, New York, 1963.
[5] B.R. Judd, *Phys. Rev.* **173**, 40 (1968).
[6] M.I. Petrashen and E.D. Trifonov, *Application of Group Theory to Quantum Mechanics*, Iliffe, London, 1969.
[7] I.G. Kaplan, *Sov. Phys.—JETP* **10**, 747 (1960).

[8] I.G. Kaplan, *Symmetry of Many-Electron Systems*, Academic Press, New York, 1975.
[9] I.G. Kaplan and U. Miranda, State-of-the-art calculations of the 3d transition-metal dimers: Mn_2 and Sc_2, in *Practical Aspects of Computational Chemistry II: An Overview of the Last Two Decades and Current Trends*, eds. J. Leszczynski and K. Shukla (Springer, Dordrecht/London, 2012), pp. 361–390.
[10] J.C. Slater, *Quantum Theory of Molecules and Solids, Vol. 1, Electronic Structure of Molecules*, McGraw-Hill, New York, 1963.
[11] I.G. Kaplan, *Sov. Phys.—JETP* **24**, 114 (1967).
[12] H. Eyring, J. Walter, and G.E. Kimball, *Quantum Chemistry*, John Wiley & Sons, Inc., New York, 1944.
[13] W. Kauzmann, *Quantum Chemistry*, Academic Press, New York, 1957.
[14] L.F. Mattheiss, *Phys. Rev.* **123**, 1209 (1961).
[15] I.G. Kaplan and O.B. Rodimova, *Sov. Phys. - JETP* **28**, 995 (1969).
[16] M. Kotani, *Proc. Phys. Math. Soc. Jpn.* **19**, 460 (1937).
[17] I.G. Kaplan and O.B. Rodimova, *Sov. Phys. - JETP* **39**, 764 (1974).
[18] I.G. Kaplan and O.B. Rodimova, *Int. J. Quantum Chem.* **10**, 699 (1976).
[19] I.G. Kaplan, *Sov. Phys. - JETP* **14**, 401 (1962).
[20] E. Wigner and E.E. Witmer, *Z. Phys.* **51**, 859 (1928).
[21] R.S. Mulliken, *Rev. Mod. Phys.* **4**, 1 (1932).
[22] G. Herzberg, *Molecular Spectra and Molecular Structure, I. Spectra of Diatomic Molecules*, D. Van Nostrand, Princeton, New Jersey/Toronto/London, 1950.
[23] D.E. Littlewood, *The Theory of Group Characters and Matrix Representations of Groups*, Oxford University Press (Clarendon), London, 1940.
[24] I.G. Kaplan and O.B. Rodimova, *Sov. Phys. Solid State* **16**, 1981 (1975).

5

Parastatistics, Fractional Statistics, and Statistics of Quasiparticles of Different Kind

5.1 Short Account of Parastatistics

As was discussed in Section 1.2, Green [1] introduced in 1953 a generalized method of second quantization with trilinear commutation relations, which include the boson and fermion commutation relations as particular cases. Namely,

$$\left[\left[a_k^+, a_l\right]_{\pm}, a_m\right]_{-} = -2\delta_{km} a_l, \tag{5.1}$$

$$\left[\left[a_k, a_l\right]_{\pm}, a_m^+\right]_{-} = 2\delta_{lm} a_k \pm 2\delta_{km} a_l, \tag{5.2}$$

$$\left[\left[a_k, a_l\right]_{\pm}, a_m\right]_{-} = 0, \tag{5.3}$$

upper sign for parabosons and lower sign for parafermions. In Eqs. (5.1)–(5.3), a_k^+ and a_k are the creation and annihilation operators for paraparticles, respectively (the second quantization formalism for bosons and fermions is described in Appendix E).

Eq. (5.1)–(5.3) are also valid, if we take the adjoint of both sides of these equations. Thus, three other parastatistical equations can be obtained. Namely:

$$\left[\left[a_k^+, a_l\right]_{\pm}, a_m^+\right]_{-} = \mp 2\delta_{lm} a_k^+, \tag{5.1a}$$

The Pauli Exclusion Principle: Origin, Verifications, and Applications, First Edition. Ilya G. Kaplan.

$$[[a_k^+, a_l^+]_\pm, a_m]_- = 2\delta_{lm} a_k^+ \pm 2\delta_{km} a_l^+, \tag{5.2a}$$

$$[[a_k^+, a_l^+]_\pm, a_m^+]_- = 0. \tag{5.3a}$$

Green found an infinite set of solutions, labeled by the integers $p(p = 1, 2, \ldots, \infty)$,

$$a_k = \sum_{\alpha=1}^{p} b_k^{(\alpha)}. \tag{5.4}$$

It is the *Green Ansatz*. For parabosons:

$$\alpha = \beta \quad \left[b_k^{(\alpha)}, b_\ell^{(\alpha)+}\right]_- = \delta_{k\ell}, \quad \left[b_k^{(\alpha)+}, b_\ell^{(\alpha)+}\right]_- = \left[b_k^{(\alpha)}, b_\ell^{(\alpha)}\right]_- = 0, \text{ as for bosons;} \tag{5.5}$$

$$\alpha \neq \beta \quad \left[b_k^{(\alpha)}, b_\ell^{(\beta)+}\right]_+ = \left[b_k^{(\alpha)+}, b_\ell^{(\beta)+}\right]_+ = \left[b_k^{(\alpha)}, b_\ell^{(\beta)}\right]_+ = 0, \text{ as for fermions.} \tag{5.6}$$

For parafermions:

$$\alpha = \beta \quad \left[b_k^{(\alpha)}, b_\ell^{(\alpha)+}\right]_+ = \delta_{k\ell}, \quad \left[b_k^{(\alpha)+}, b_\ell^{(\alpha)+}\right]_+ = \left[b_k^{(\alpha)}, b_\ell^{(\alpha)}\right]_+ = 0, \text{ as for fermions;} \tag{5.7}$$

$$\alpha \neq \beta \quad \left[b_k^{(\alpha)}, b_\ell^{(\beta)+}\right]_- = \left[b_k^{(\alpha)+}, b_\ell^{(\beta)+}\right]_- = \left[b_k^{(\alpha)}, b_\ell^{(\beta)}\right]_- = 0, \text{ as for bosons.} \tag{5.8}$$

The value of p in the Green Ansatz is called the *rank of parastatistics*. As follows from Eqs. (5.5) and (5.7), the solution of the Green Ansatz for $p = 1$ (when only one type of operators b exists) reduces to the usual boson and fermion operators in paraboson and parafermion cases, respectively. It is worthwhile to note that such mixed boson–fermion behavior of the operators $b_k^{(\alpha)}$ is well known in solid-state physics and called as *paulionic*, see the book by Davydov [2].

The application of operators (5.4) to the vacuum state is found using Eqs. (5.5)–(5.8). So,

$$a_k|0\rangle = 0, \quad a_k a_\ell^+|0\rangle = p\delta_{k\ell}|0\rangle \quad \text{for all } k, \ell. \tag{5.9}$$

For the parafermi statistics of order p, no more than p parafermions can be in the same quantum state; thus the maximum occupation numbers are equal to the rank of parafermi statistics

$$\left(a_k^+\right)^N|0\rangle \neq 0 \quad \text{for} \quad N \leq p, \tag{5.10}$$

$$\left(a_k^+\right)^{p+1}|0\rangle = 0. \tag{5.11}$$

As follows from Eq. (5.7), the parafermi statistics at $p = 1$ is reduced to the Fermi–Dirac statistics. The similar property is valid for the parabose statistics, which at $p = 1$ is reduced to the Bose–Einstein statistics. Although in the case of the parabose statistics there is no restriction in the occupation numbers at any p.

The particle number operator in parastatistics is defined as

$$\hat{N}_k = \frac{1}{2}\left[a_k^+, a_k\right]_\pm \mp \frac{1}{2}p. \tag{5.12}$$

It satisfies the general property of the particle number operators, cf. Appendix E, where Eq. (E.15) is written for $k = \ell$,

$$\left[\hat{N}_k, a_\ell^+\right]_- = \delta_{k\ell} a_\ell^+. \tag{5.13}$$

Let us show that in the Bose (Fermi) case the expression (5.12) turns to a well-known expression for the particle number operator in the second quantization formalism, see Eq. (E.14) for bosons and Eq. (E.14a) for fermions.

In the paraboson case and $p = 1$, Eq. (5.12) can be written as

$$\hat{N}_k = \frac{1}{2}\left(a_k^+ a_k + a_k a_k^+ - 1\right). \tag{5.14}$$

From the boson commutation relation (E.11) it follows that $a_k a_k^+ = 1 + a_k^+ a_k$. Substituting this expression into Eq. (5.14) we obtain

$$\hat{N}_k = a_k^+ a_k. \tag{5.15}$$

The similar expression is obtained also in the parafermi case for $p = 1$.

As was shown in Refs. [3, 4], the parafermion and paraboson algebras are the Lie algebras of the orthogonal and symplectic groups, respectively. For example, the algebra for ν parafermion operators a_r and their adjoints a_r^+ $(r = 1, 2, \ldots, \nu)$, which constitutes a parafermi ring, is the Lie algebra of the orthogonal group $SO_{2\nu+1}$ in $2\nu + 1$ dimensions; more details about the parastatistics is presented in the book by Ohnuki and Kamefuchi [5].

In spite of numerous studies of all known elementary particles, the elementary particles obeying the parastatistics were not revealed. On the other hand, as has been discussed in Refs. [6–8], the ordinary fermions, which differ by some internal quantum numbers, but are similar dynamically, can be described by the parafermi statistics. In this case, fermions with different internal quantum numbers are considered as different particles or as different states of dynamically equivalent particles. The parafermi statistics of rank p describes systems with p different types of fermions. So *quarks* with three colors obey the parafermi statistics of rank $p=3$; *nucleons* in nuclei (isotope spin 1/2) obey the parafermi statistics of rank $p=2$. The total wave function for such parafermions always can be constructed as an antisymmetric function in full accordance with the Pauli exclusion principle.

Although, all elementary particles known at present are bosons or fermions, the parastatistics can be realized for quasiparticles. As was shown in 1976 by Kaplan [9], the quasiparticles in a periodical lattice (the Frenkel excitons and magnons) obeyed a modified parafermi statistics of rank M, where M is the number of equivalent lattice sites within the delocalization region of collective excitations. Later on, it was shown that the introduced by Kaplan modified parafermi statistics [9] is valid for different types of quasiparticles in a periodical lattice: polaritons [10, 11], defectons in quantum crystals [12], the Wannier–Mott excitons [13], delocalized holes in crystals [14], and delocalized coupled hole pairs [15]. In next section we will discuss in detail the statistics and properties of systems of noninteracting holes in a periodical lattice, following the approach developed in Ref. [9].

5.2 Statistics of Quasiparticles in a Periodical Lattice

5.2.1 Holes as Collective States

When a hole (a positive charge) is created in some atom in a lattice or in a monomer in a polymer, it can migrate in a lattice or along a polymer chain. There is an equal probability for the location of the hole on each site of the same nature. As a result, the collective state is formed. This collective state can be considered as a charge wave or as a quasiparticle. Below we will study the statistics of such quasiparticles at arbitrary concentrations. For this purpose, we use the second quantization formalism, see Appendix E.

Usually holes are considered as fermion particles with spin 1/2 and positive charge. In the second quantization formalism, the hole operators are adjoint to the electron operators and obey the fermion commutation relations. Although we should take into account that in real systems holes are located on many-electron atoms or molecules and can have different values of spin S. For example, holes in the CuO_2 planes in high T_c cuprate oxides have $S=0$ [16]. As we mentioned above, in a periodical lattice holes are delocalized and form a collective state.

In the following discussion we consider holes as positive charged atoms (molecules) with $S=0$, or as spinless quasiparticles. In the absence of dynamical interaction between holes, the model Hamiltonian for an arbitrary lattice with one type of noninteracting holes can be written as

$$H = \varepsilon_0 \sum_n b_n^+ b_n + \sum_{n,\, n'} M_{nn'} b_n^+ b_{n'}, \tag{5.16}$$

where b_n^+ and b_n are the hole creation and annihilation operators, respectively; ε_0 is the energy of the hole creation in a lattice; and $M_{nn'}$ is the so-called hopping integral, characterizing the efficiency of charge transfer (hopping) from site n to site n'. The study of the hole migration process has a great importance for elucidating the mechanism of high T_c superconductivity [17]. The efficiency of charge transfer also determines the properties of organic conductors and semiconductors, and irradiated polymers [18].

According to the theory of the resonance interaction (see section 2.2 in book [19]), the hopping integral can be expressed as the resonance integral in the following form:

$$M_{nn'} = \left\langle \Psi_0\left(A_{n'}^+\right)\Psi_0(A_n)\middle| V_{\text{int}} \middle| \Psi_0(A_{n'})\Psi_0\left(A_n^+\right)\right\rangle, \tag{5.17}$$

where $\Psi_0(A_n)$ and $\Psi_0\left(A_n^+\right)$ are the ground-state many-electron wave functions for neutral and ionized atoms (monomers) located at site n. In the one-electron approximation the expression (5.17) has the same physical sense as the hopping integral in the Hubbard Hamiltonian [20].

If the wave functions in Eq. (5.17) are not overlapping, the operators $b_{n'}^+$ and b_n acting on different sites must commute. Thus, they obey the Bose commutation relations:

$$\left[b_n, b_{n'}^+\right]_- = [b_n, b_{n'}]_- = \left[b_n^+, b_{n'}^+\right]_- = 0 \quad \text{for } n \neq n'. \tag{5.18}$$

In the Hamiltonian given by Eq. (5.16) is assumed that on a site can be created only one hole and only of one type. Although, the definition of the hole is quite general and it can correspond to an arbitrary n-fold ionized state, we do not consider states with two holes on one site. From this it follows that

$$\left(b_n^+\right)^2 |0\rangle = 0, \tag{5.19}$$

where $|0\rangle$ is the vacuum state. As the vacuum state, we consider a state in which all sites of a lattice are neutral. The operators acting on one site satisfy the Fermi commutation relations:

$$\begin{aligned} &\left[b_n, b_n^+\right]_+ = 1, \\ &[b_n, b_n]_+ = \left[b_n^+, b_n^+\right]_+ = 0. \end{aligned} \tag{5.20}$$

On the other hand, according to Eq. (5.18), the hole operators acting on different sites commute. The operators with the commutation relations (5.18)–(5.20) are called the Pauli operators [2] and describe the paulion particles. Thus, the introduced operators for holes are paulions as the second quantization operators for spin [21] or operators for electronic excitations in crystals [2].

The Hamiltonian (5.16) can be diagonalized by some unitary transformation:

$$B_{\mathbf{q}} = \frac{1}{\sqrt{M}}\sum_{n} u_{\mathbf{q}n} b_n; \qquad B_{\mathbf{q}}^{+} = \frac{1}{\sqrt{M}}\sum_{n} u_{\mathbf{q}n}^{*} b_n^{+}, \tag{5.21}$$

where M is the number of lattice sites on which the hole can be created. With the new operators, the Hamiltonian is transformed from the site representation to the quasi-momentum representation in which it has the diagonalized form

$$H = \sum_{\mathbf{q}} \varepsilon_{\mathbf{q}} B_{\mathbf{q}}^{+} B_{\mathbf{q}}. \tag{5.22}$$

The action of the operator $B_{\mathbf{q}}^{+}$ on the vacuum state $|0\rangle$ creates a collective charge state with a hole distributed among all equivalent sites. This state can be considered as a quasiparticle similar to exciton [2] or magnon [21] quasiparticles; so, it is natural to call it *holon*. For lattice with one atom (molecule) per cell, the energy of holon is given by the following expression:

$$\varepsilon_{\mathbf{q}} = \varepsilon_0 + \sum_{n'(\neq n)} M_{nn'} \exp[i\mathbf{q}(\mathbf{r}_n - \mathbf{r}_{n'})]. \tag{5.23}$$

Let us note that the collective charge state, which we named as holon, is different from the Anderson holon introduced in his resonating valence bond (RVB) model of high-temperature superconductivity [22–24]. In the Anderson model, the electronic excitation spectrum is presented as two separated branches: charge spinless holons (bosons) and chargeless spinons (fermions), corresponding to charge and spin degrees of freedom of electrons. As we will demonstrate further, the holons, discussed in this section, are bosons only at a low concentration limit, while at large concentrations they obey the modified parafermi statistics.

5.2.2 *Statistics and Some Properties of Holon Gas*

The holon operators $B_{\mathbf{q}}$ and $B_{\mathbf{q}}^{+}$ are connected with the hole operators b_n andb_n^{+} by unitary transformation (5.21). Since the hole operators do not obey neither the boson nor the fermion commutation relations, the unitary transformation in the general case is not canonical; it means, it does not preserve the commutation properties of the operators transformed. In fact, using the commutation relations (5.18)–(5.20), the following commutators hold:

$$\left[B_{\mathbf{q}}, B_{\mathbf{q}'}^{+}\right]_{-} = M^{-1}\sum_{n} u_{\mathbf{q}n}u_{\mathbf{q}'n}^{*}\left(1-2b_{n}^{+}b_{n}\right),$$
$$\left[B_{\mathbf{q}}, B_{\mathbf{q}'}^{+}\right]_{+} = M^{-1}\sum_{n} u_{\mathbf{q}n}u_{\mathbf{q}'n}^{*} + 2M^{-1}{\sum}'_{n,n'} u_{\mathbf{q}n}u_{\mathbf{q}'n'}^{*}b_{n'}^{+}b_{n}, \tag{5.24}$$

where the prime at the sum means that $n \neq n'$.

For a simple periodical lattice with one atom (molecule) per cell, the unitary transformation that diagonalizes the Hamiltonian is completely determined by the translational symmetry of the lattice, in such case the $u_{\mathbf{q}n}$ in Eq. (5.21) is given by a simple exponent

$$u_{\mathbf{q}n} = \exp(-i\mathbf{q}\cdot\mathbf{r}_{n}),$$

and Eqs. (5.24) are represented as

$$\left[B_{\mathbf{q}}, B_{\mathbf{q}'}^{+}\right]_{-} = \delta_{\mathbf{q}\mathbf{q}'} - \frac{2}{M}\sum_{n} \exp\{i(\mathbf{q}'-\mathbf{q})\cdot\mathbf{r}_{n}\}b_{n}^{+}b_{n}, \tag{5.25}$$

$$\left[B_{\mathbf{q}}, B_{\mathbf{q}'}^{+}\right]_{+} = \delta_{\mathbf{q}\mathbf{q}'} + \frac{2}{M}{\sum_{n,n'}}' \exp\{i(\mathbf{q}'\cdot\mathbf{r}_{n'}-\mathbf{q}\cdot\mathbf{r}_{n})\}b_{n'}^{+}b_{n}. \tag{5.25a}$$

Let us find for the holon operators $B_{\mathbf{q}}$ and $B_{\mathbf{q}}^{+}$ the trilinear commutators as in the parastatistics. The sums on the right-hand side of commutators (5.25), which contains the hole operators b_{n}^{+}, b_{n}, disappear and the commutation relations will not contain other kinds of operators. For the parafermi case we obtain

$$\left[\left[B_{\mathbf{q}}^{+}, B_{\mathbf{q}'}\right]_{-}, B_{\mathbf{q}''}\right]_{-} = -2M^{-1}B_{\bar{\mathbf{q}}};\;\; \bar{\mathbf{q}} = \mathbf{q}'+\mathbf{q}''-\mathbf{q}, \tag{5.26}$$

$$\left[\left[B_{\mathbf{q}}^{+}, B_{\mathbf{q}'}\right]_{-}, B_{\mathbf{q}''}^{+}\right]_{-} = 2M^{-1}B_{\bar{\mathbf{q}}}^{+};\;\; \bar{\mathbf{q}} = \mathbf{q}-\mathbf{q}'+\mathbf{q}''. \tag{5.26a}$$

The relations (5.26) for $\mathbf{q}=\mathbf{q}''$ and (5.26a) for $\mathbf{q}'=\mathbf{q}''$ are, within a normalization factor M, identical with the corresponding commutation relations for parafermions (see Eqs. 5.1 and 5.1a, respectively). But for arbitrary $\mathbf{q}$, $\mathbf{q}'$, and $\mathbf{q}''$ there is one essential difference with the parafermion relations. In the latter there is the Kronecker symbol whereas in Eq. (5.26) the Kronecker symbol is absent, the state vector $\bar{\mathbf{q}}$ in the right-hand side of these equations is determined by conservation law of the quasimomentum in a periodical lattice. It follows from the physics of the considered system. On the one hand, it is natural; however, on the other hand, this difference with the parafermi commutation relations leads to important physical consequences which we will discuss below.

To determine the action of the operator $B_{\mathbf{q}}$ on the vacuum state $|0\rangle$, which is the ground state at the Fermi level, it suffices to find the action of the Pauli operators on this state. The Pauli operators satisfy the following natural conditions:

$$b_n|0\rangle = 0, \quad b_{n'}b_n^+|0\rangle = \delta_{nn'}|0\rangle. \tag{5.27}$$

The same result is valid for the holon operators:

$$B_{\mathbf{q}}|0\rangle = 0, \quad B_{\mathbf{q}'}B_{\mathbf{q}}^+|0\rangle = \delta_{\mathbf{q}\mathbf{q}'}|0\rangle. \tag{5.28}$$

Using the definition of the operators $B_{\mathbf{q}}^+$, Eq. (5.21), and the commutation relations (5.18)–(5.20) for b_n^+, we obtain for the two-particle case

$$\left(B_{\mathbf{q}}^+\right)^2|0\rangle = 2M^{-1}\sum_{n<n'} u_{\mathbf{q}n'}^* u_{\mathbf{q}n}^* b_{n'}^+ b_n^+|0\rangle, \tag{5.29}$$

for the general case $N < M$

$$\left(B_{\mathbf{q}}^+\right)^N|0\rangle = \left(\frac{N!}{M^{N/2}}\right)\sum_{n_1<n_2\cdots<n_N} u_{\mathbf{q}n_N}^* \ldots u_{\mathbf{q}n_1}^* b_{n_N}^+ \ldots b_{n_1}^+|0\rangle, \tag{5.30}$$

and for the maximum value $N = M$

$$\left(B_{\mathbf{q}}^+\right)^M|0\rangle = \left(\frac{M!}{M^{M/2}}\right) u_{\mathbf{q}n_M}^* \ldots u_{\mathbf{q}n_1}^* b_{n_M}^+ \ldots b_{n_1}^+|0\rangle. \tag{5.31}$$

In the case $N = M + 1$ we always will have two particles in one site; therefore, using Eq. (5.19), we obtain

$$\left(B_{\mathbf{q}}^+\right)^{M+1}|0\rangle = 0. \tag{5.32}$$

Thus, one state can be occupied by up to M quasiparticles. It means that holons satisfy some modified parafermi statistics of rank M. It is important to stress that this conclusion does not depend on the special choice of the unitary transformation and is valid also for quasiparticles in a complicated lattice with several atoms (molecules) per cell.

A function describing a state of N noninteracting particles, each with energy $\varepsilon_{\mathbf{q}}$, is given by the usual equation

$$|N_{\mathbf{q}}\rangle = C_N\left(B_{\mathbf{q}}^+\right)^N|0\rangle. \tag{5.33}$$

To find the expression for the normalization factor it is convenient to use the following operator equation, which is obtained from Eq. (5.26a):

$$B_{\mathbf{q}}B_{\mathbf{q}'}^{+}B_{\mathbf{q}''}^{+}=B_{\mathbf{q}'}^{+}B_{\mathbf{q}}B_{\mathbf{q}''}^{+}+B_{\mathbf{q}''}^{+}B_{\mathbf{q}}B_{\mathbf{q}'}^{+}-B_{\mathbf{q}''}^{+}B_{\mathbf{q}'}^{+}B_{\mathbf{q}}-\frac{2}{M}B_{\bar{\mathbf{q}}}^{+};\ \bar{\mathbf{q}}=\mathbf{q}'+\mathbf{q}''-\mathbf{q},\tag{5.34}$$

or following from it

$$B_{\mathbf{q}'}\left(B_{\mathbf{q}}^{+}\right)^{2}=2B_{\mathbf{q}}^{+}B_{\mathbf{q}'}B_{\mathbf{q}}^{+}-\left(B_{\mathbf{q}}^{+}\right)^{2}B_{\mathbf{q}'}-\frac{2}{M}B_{\bar{\mathbf{q}}}^{+};\ \bar{\mathbf{q}}=2\mathbf{q}-\mathbf{q}'\tag{5.35}$$

Using Eq. (5.35), we find by the induction method the expression for the normalization factor, which differs from that of a Bose system:

$$C_N=\left[N!\left(1-\frac{1}{M}\right)\left(1-\frac{2}{M}\right)\cdots\left(1-\frac{N-1}{M}\right)\right]^{-\frac{1}{2}}.\tag{5.36}$$

The results of applying the operators $B_{\mathbf{q}}^{+}$ and $B_{\mathbf{q}}$ to the state vector $|N_{\mathbf{q}}\rangle=C_{N_{\mathbf{q}}}\left(B_{\mathbf{q}}^{+}\right)^{N_{\mathbf{q}}}|0\rangle$ are

$$B_{\mathbf{q}}^{+}|N_{\mathbf{q}}\rangle=\sqrt{(N_{\mathbf{q}}+1)\left(1-\frac{N_{\mathbf{q}}}{M}\right)}|N_{\mathbf{q}}+1\rangle\tag{5.37}$$

$$B_{\mathbf{q}}|N_{\mathbf{q}}\rangle=\sqrt{N_{\mathbf{q}}\left(1-\frac{N_{\mathbf{q}}-1}{M}\right)}|N_{\mathbf{q}}-1\rangle.\tag{5.38}$$

Eq. (5.37) shows that the effect of applying $B_{\mathbf{q}}^{+}$ on a state with a maximum occupation number $N_{\mathbf{q}}=M$ is equal to zero. As $M\to\infty$, relations (5.37) and (5.38) turn into the well-known relations for bosons, see Eqs. (E.7) and (E.3).

From Eqs. (5.37) and (5.38), it follows that

$$B_{\mathbf{q}}^{+}B_{\mathbf{q}}|N_{\mathbf{q}}\rangle=N_{\mathbf{q}}\left(1-\frac{N_{\mathbf{q}}-1}{M}\right)|N_{\mathbf{q}}\rangle.\tag{5.39}$$

This means that the operator $B_{\mathbf{q}}^{+}B_{\mathbf{q}}$ is not a particle number operator as it is in the case of boson, fermion, and paulion operators. For the commutator we obtain

$$\left[B_{\mathbf{q}},B_{\mathbf{q}}^{+}\right]_{-}|N_{\mathbf{q}}\rangle=\left(1-\frac{2N_{\mathbf{q}}}{M}\right)|N_{\mathbf{q}}\rangle.\tag{5.40}$$

Hence,

$$\frac{1}{2}M\left(1-\left[B_{\mathbf{q}},B_{\mathbf{q}}^{+}\right]_{-}\right)|N_{\mathbf{q}}\rangle=N_{\mathbf{q}}|N_{\mathbf{q}}\rangle, \tag{5.41}$$

and the expression for the operator of the particle number, $\hat{N}_{\mathbf{q}}$, in state $\mathbf{q}$ is

$$\hat{N}_{\mathbf{q}}=\frac{1}{2}M\left(1-\left[B_{\mathbf{q}},B_{\mathbf{q}}^{+}\right]_{-}\right). \tag{5.42}$$

The trilinear commutation relation (5.26a) can be expressed via the $\hat{N}_{\mathbf{q}}$ as

$$\left[\hat{N}_{\mathbf{q}},B_{\mathbf{q}'}^{+}\right]_{-}=B_{\mathbf{q}'}^{+}, \tag{5.43}$$

while the parafermionic commutation relation, because of the Kronecker symbol, gives for the particle number operator, instead of Eq. (5.43), the relation

$$\left[\hat{N}_{\mathbf{q}},B_{\mathbf{q}}^{+}\right]_{-}=B_{\mathbf{q}}^{+}. \tag{5.44}$$

Eq. (5.44) is really the commutation relation which must be fulfilled by the operator of particle number, since the action of the commutator (5.44) on the state $|N_{\mathbf{q}}\rangle$ is equivalent to the increase of the eigenvalue of the operator $\hat{N}_{\mathbf{q}}$, cf. also Eq. (E.15) in Appendix E. But the commutation relation (5.43) leads to an unusual result: the eigenvalue of the operator $\hat{N}_{\mathbf{q}}$ depends upon the occupation number of state $\mathbf{q}'$ for all states $\mathbf{q}'$. Thus, the operator $\hat{N}_{\mathbf{q}}$ (5.42) cannot belong only to the state $\mathbf{q}$ and have to be considered as the operator of the total number of quasiparticles. It is easy to show that this is indeed the case.

Substituting Eq. (5.25) into the expression (5.42) for $\hat{N}_{\mathbf{q}}$ we obtain

$$\hat{N}_{\mathbf{q}}=\frac{1}{2}M\left(1-\left[B_{\mathbf{q}},B_{\mathbf{q}}^{+}\right]_{-}\right)=\sum_{n}b_{n}^{+}b_{n}=\hat{N}. \tag{5.45}$$

Since the number of quasiparticles is equal to the number of holes at the sites, the operator (5.42) is the operator of the total number of quasiparticles, and it does not depend on $\mathbf{q}$. From Eq. (5.45) it follows that

$$\left[B_{\mathbf{q}},B_{\mathbf{q}}^{+}\right]_{-}=1-\frac{2\hat{N}}{M}. \tag{5.46}$$

This means that for $\langle\hat{N}\rangle\ll M$ the quasiparticles satisfy the Bose statistics with good accuracy.

The Hamiltonian of an ideal gas of holons must be linear in the particle number operator, cf. Eq. (E.23):

$$H=\sum_{\mathbf{q}}\varepsilon_{\mathbf{q}}\hat{N}_{\mathbf{q}}. \tag{5.47}$$

However, as was shown above, we cannot introduce the holon number operator for a particular state because of the immanent coupling of different holon states even in the absence of dynamical interaction. Thus, it is impossible to describe an ideal gas of holons beyond the Bose approximation, which is valid only for small concentration of holons. In general, the interaction between quasiparticles is always present. This kind of interaction depending on the deviation of quasiparticle statistics from the Bose (Fermi) statistics is called, after Dyson [25], the *kinematic* interaction.

In order to estimate the magnitude of the kinematic interaction, it suffices to calculate the expectation value of the Hamiltonian (5.22) with the state vector (5.33),

$$E_{N_{\mathbf{q}}}=\langle N_{\mathbf{q}}|H|N_{\mathbf{q}}\rangle=C_N^2\left\langle 0\left|(B_{\mathbf{q}})^N\sum_{\mathbf{q}'}\varepsilon_{\mathbf{q}'}B_{\mathbf{q}'}^{+}B_{\mathbf{q}'}\left(B_{\mathbf{q}}^{+}\right)^N\right|0\right\rangle. \tag{5.48}$$

To calculate this mean value it is convenient to use the operator equation (5.34) and the properties (5.28). As a result, for two quasiparticles with energy $\varepsilon_{\mathbf{q}}$, we obtain

$$E_{2_{\mathbf{q}}}=2\varepsilon_{\mathbf{q}}\left(1-\frac{1}{M}\right)+\frac{2}{M-1}\frac{1}{M}\sum_{\mathbf{q}'(\neq\mathbf{q})}\varepsilon_{\mathbf{q}'}, \tag{5.49}$$

and for the general case of N quasiparticles

$$\begin{aligned}E_{N_{\mathbf{q}}}&=N\varepsilon_{\mathbf{q}}\left(1-\frac{N-1}{M}\right)+\frac{N}{M-1}\frac{N-1}{M}\sum_{\mathbf{q}'(\neq\mathbf{q})}\varepsilon_{\mathbf{q}'}\\&=N\left[\varepsilon_{\mathbf{q}}+\frac{N-1}{M}\left(\frac{1}{M-1}\left(\sum_{\mathbf{q}'(\neq\mathbf{q})}\varepsilon_{\mathbf{q}'}\right)-\varepsilon_{\mathbf{q}}\right)\right].\end{aligned} \tag{5.50}$$

The second terms in Eq. (5.50) contain the concentration of quasiparticles.

Eq. (5.50) can be written as

$$E_{N_{\mathbf{q}}}=N\left[\varepsilon_{\mathbf{q}}+\frac{N-1}{M}\left(\bar{\varepsilon}-\varepsilon_{\mathbf{q}}\right)\right], \tag{5.51}$$

where

$$\bar{\varepsilon}=\frac{1}{M-1}\sum_{\mathbf{q}'(\neq\mathbf{q})}\varepsilon_{\mathbf{q}'} \tag{5.52}$$

denotes the mean energy of the holon band. In Eq. (5.51) the second term describes the kinematic interaction, so the holon gas gets a more ideal behavior when the energy $\varepsilon_{\mathbf{q}}$ comes nearer to the mean energy $\bar{\varepsilon}$ of the band.

The number of holons is equal to the number of created holes. The latter cannot exceed the number of lattice sites M on which the hole can be created. So, all created holons can occupy one state, for example, the ground state. It means that, in spite of nonbosonic behavior of the holon gas, there is no statistical prohibition on the Bose–Einstein condensation phenomenon in holon systems. On the other hand, the holon gas is always nonideal (because of the kinematic interaction). The study on the Bose–Einstein condensation in nonideal systems requires a special treatment of the stability of the Bose condensate [26]. For rigorous study of this problem we must include also a dynamic interaction and consider interplay between kinematic and dynamic interactions in the holon system, as we had done for the molecular exciton system in Ref. [27].

The method developed for holons can be extended to study the system of coupled holes in high-T_c superconducting ceramics [15]. In this case it is convenient to use the Hubbard model with the coupled interaction term, see next subsection.

5.2.3 *Statistics of Hole Pairs*

At present, it is well established that the conductivity in high-T_c ceramics has a hole origin with charge of carriers equal to $+2e$. Here, we present the results of our study of statistics and some physical properties of the hole-pair system.

In the second quantization formalism in the site representation, the model Hamiltonian for one type of spinless holes is

$$H=\varepsilon_0\sum_n b_n^+ b_n+\sum_{nn'} M_{nn'} b_n^+ b_{n'}+\sum_{nn'} V_{nn'} b_n^+ b_{n'}^+ b_{n'} b_n, \tag{5.53}$$

where ε_0 is the energy for the hole creation in a lattice, $M_{nn'}$ is the so-called hopping integral, see discussion of Eq. (5.17), and $V_{nn'}$ is the hole–hole interaction term, it does not depend explicitly on the nature of interaction between holes since it is included as a parameter in the Hamiltonian.

As we showed in previous section, the hole creation, b_n^+, and hole annihilation, b_n, operators are characterized by the paulion properties, Eqs. (5.18)–(5.20), that is,

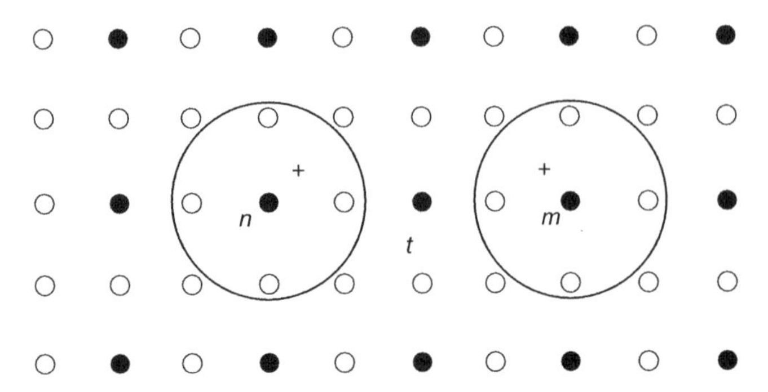

Fig. 5.1 Schematic representation of a coupled hole pair on the CuO_2 plane in high-T_c superconducting ceramics; black circles are Cu and open circles are O

the operators acting on different sites obey the boson commutation relations, while the operators acting on one site obey the fermion commutation relations.

Suppose that the hole–hole interaction term in (5.53) is attractive. In this case under some conditions, the coupled hole pairs can be formed. The operators of creation and annihilation of the hole pair are defined as usual:

$$a_t^+ = b_n^+ b_m^+, \qquad a_t = b_m b_n. \tag{5.54}$$

where t denotes the localization point of the center of mass of the coupled hole pair, see Fig. 5.1. In high-T_c ceramics, the hole-pair localization region is not large: the correlation length in the CuO_2 planes is of the order of (10–12) Å. It is easy to verify that the hole-pair operators a_t^+ and a_t obey the same paulion commutation relations as the hole operators.

Let us assume that all hole pairs have the same size, and the region of the hole pair localization can be repeated in crystal so that the points t form a "superlattice" with M sites. We will show below that in this case and only in this case the coupled hole pairs in the quasi-momentum space obey, as holons, the modified parafermi statistics with the value of the quasi-momentum defined by the quasi-momentum conservation law.

The model Hamiltonian for hole pairs can be presented as

$$H = \sum_t \varepsilon_p a_t^+ a_t + \sum_{tt'} M_{tt'} a_t^+ a_{t'}, \tag{5.55}$$

where $\varepsilon_p = 2\varepsilon_0 + V_0$ is the energy of the coupled hole pair, V_0 is the attractive potential between holes, which we assume to be the same for all pairs, as in the BCS approach, and $M_{tt'}$ is the hopping integral for a hole pair moving as a whole entity. The Hamiltonian (5.55) can be transformed by some unitary transformation:

$$A_{\mathbf{q}} = \frac{1}{\sqrt{M}} \sum_{t=1}^{M} u_{\mathbf{q}t} a_t, \quad A_{\mathbf{q}}^{+} = \frac{1}{\sqrt{M}} \sum_{t=1}^{M} u_{\mathbf{q}t}^{*} a_t^{+} \tag{5.56}$$

to the diagonalized form in the quasi-momentum space,

$$H = \sum_{\mathbf{q}} \varepsilon_{\mathbf{q}} A_{\mathbf{q}}^{+} A_{\mathbf{q}}. \tag{5.57}$$

If the superlattice made by points t in Fig. 5.1 is simple with one site per cell, the unitary transformation (5.56) is completely determined by the translation symmetry of the lattice and the coefficients $u_{\mathbf{q}t} = \exp(-i\mathbf{q}r_t)$. The self-energy of the diagonalized Hamiltonian (5.57) is equal to

$$\varepsilon_{\mathbf{q}} = \varepsilon_p + \sum_{t'(\neq t)} M_{tt'} \exp[i\mathbf{q} \cdot (\mathbf{r}_t - \mathbf{r}_{t'})]. \tag{5.58}$$

Since the operators (5.54) obey neither the boson nor the fermion commutation relations, the unitary transformation in general case is not canonical; that is, it does not preserve the commutation properties of the operators transformed. In particular, the operators (5.56) do not describe the paulion quasiparticles. As we showed in previous section for holes, such operators obey the modified parafermi statistics. For lattices, diagonalized by an exponential unitary transformation, the operators (5.56) obey trilinear commutation relations (5.26), which correspond to the modified parafermi statistics [9, 14]. It can be proved that the rank of parastatistics is equal to the number of sites, M, in the superlattice. This means that one state can be occupied by up to M hole pairs:

$$\left(A_{\mathbf{q}}^{+}\right)^{N} |0\rangle \neq 0, \quad N \leq M \tag{5.59}$$

$$\left(A_{\mathbf{q}}^{+}\right)^{M+1} |0\rangle = 0. \tag{5.60}$$

The state with N noninteracting pairs, each with the same $\mathbf{q}$, is defined by the expression

$$|N_{\mathbf{q}}\rangle = C_N \left(A_{\mathbf{q}}^{+}\right)^{N} |0\rangle, \tag{5.61}$$

with the normalization factor C_N given by Eq. (5.36). It is also easy to find that

$$A_{\mathbf{q}}^{+}|N_{\mathbf{q}}\rangle = \sqrt{(N_{\mathbf{q}}+1)\left(1-\frac{N_{\mathbf{q}}}{M}\right)}|N_{\mathbf{q}}+1\rangle \tag{5.62}$$

$$A_{\mathbf{q}}|N_{\mathbf{q}}\rangle = \sqrt{N_{\mathbf{q}}\left(1-\frac{N_{\mathbf{q}}-1}{M}\right)}|N_{\mathbf{q}}-1\rangle. \tag{5.63}$$

From Eq. (5.62) it follows that the result of applying A_q^+ on the state with the maximum occupation number $N_q = M$ is equal to zero. As $M \to \infty$, both relations turn into the well-known relations for bosons.

According to Eqs. (5.62) and (5.63),

$$A_{\mathbf{q}}^{+}A_{\mathbf{q}}|N_{\mathbf{q}}\rangle = N_{\mathbf{q}}\left(1-\frac{N_{\mathbf{q}}-1}{M}\right)|N_{\mathbf{q}}\rangle \tag{5.64}$$

Thus, similar to the holon case, the operator $A_{\mathbf{q}}^{+}A_{\mathbf{q}}$ is not a particle number operator in a state **q**. As was demonstrated in the previous section, for the modified parafermi statistics the operator of particle number in a state **q** does not exist. What can be defined is the operator of the total number of hole pairs, $\hat{N}$. For the commutator, the following relation is valid:

$$\left[A_{\mathbf{q}}, A_{\mathbf{q}}^{+}\right]_{-} = 1 - \frac{2\hat{N}}{M}. \tag{5.65}$$

Only for small concentrations when $\langle \hat{N} \rangle / M \ll 1$, the hole pairs satisfy the Bose statistics.

Since the operator $A_{\mathbf{q}}^{+}A_{\mathbf{q}}$ is not the hole-pair number operator, the diagonalized Hamiltonian (5.57) does not describe the ideal gas of the hole pairs. For same reasons, as were discussed in Section 5.2.2, the latter does not exist in principle. Even in the absence of dynamic interactions, some immanent interaction in the hole-pair system, called the *kinematic* interaction [25], is always present; its origin is in the deviation of the hole-pair statistics from the Bose (Fermi) statistics. The expression for the kinematic interaction is given by the second term in Eq. (5.51).

Thus, there is an immanent coupling among all states of the hole pair band. Therefore, we cannot define the independent quasiparticles in some particular state. As we mentioned above, the ideal gas of the hole pairs does not exist fundamentally. It can exist only in the low concentration limit, in which the kinematic interaction becomes small, and we get the case of the Bose statistics, as follows from Eq. (5.65).

In high-T_c superconducting ceramics, the maximum T_c is achieved for a hole concentration in CuO_2 planes equal to $0.2-0.25$ per CuO_2 unit [28, 29]. The same order of magnitude has to be for the hole-pair concentration because the latter is counted not per CuO_2 units, but per the number of sites M in the superlattice. Thus, the deviations from the Bose statistics for the hole-pair system are not negligible and have to be taken into account.

As we discussed above, the hole pairs obey the modified parafermi statistics of rank M, so, one state can be occupied by up to M hole pairs. The number of hole pairs cannot exceed the number of sites M in the superlattice. This means that, in spite of the non-boson behavior of the hole-pair system, there is no statistical prohibition for the Bose–Einstein condensation. On the other hand, the hole-pair system is always nonideal (due to the kinematic interaction). For a rigorous study of the Bose–Einstein condensation phenomenon, we have to include also a dynamic interaction and consider an interplay between kinematic and dynamic interactions to study the stability of the Bose condensate, as it was done for the molecular exciton system in Ref. [27].

5.3 Statistics of Cooper's Pairs

As we discussed in Section 1.2, the operators of creation, b_k^+, and annihilation, b_k, of Cooper's pair in a state $(\boldsymbol{k}\alpha, -\boldsymbol{k}\beta)$, where $\boldsymbol{k}$ is the electron momentum, α and β are the spin projections, are defined as products of the fermionic creation, $c_{k\alpha}^+$, and annihilation, $c_{k\alpha}$, operators, satisfying the fermion commutation relations,

$$b_k^+ = c_{k\alpha}^+ c_{-k\beta}^+, \quad b_k = c_{-k\beta} c_{k\alpha}. \tag{5.66}$$

Let us call these operators, following Schrieffer [30], as *pairon* operators.

The Cooper pairs have spin $S = 0$, so the permutation symmetry of their wave functions is bosonic, they are symmetric. But their operators do not obey the boson commutation relations. The direct calculation gives

$$\left[b_k, b_{k'}^+\right]_- = \delta_{kk'}\left(1 - c_{k\alpha}^+ c_{k\alpha} - c_{k\beta}^+ c_{k\beta}\right). \tag{5.67}$$

Only in the case $k \neq k'$,

$$\left[b_k, b_{k'}\right]_- = 0 \tag{5.68}$$

and Cooper's pairs obey the boson commutation relations. For $k = k'$

$$\left[b_k, b_k^+\right]_- = 1 - c_{k\alpha}^+ c_{k\alpha} - c_{k\beta}^+ c_{k\beta}. \tag{5.69}$$

Due to the fermion nature of electrons constituting Cooper's pairs, the commutation relations for the Cooper pair operators are not bosonic and contain the fermionic operators. For the same reason Cooper's pairs have the fermionic occupation numbers,

$$\left(b_k^+\right)^2 = (b_k)^2 = 0 \tag{5.70}$$

It would be more natural, if the commutation relations for the pairon operators (5.67) do not have to include other kinds of operators. One of the ways to achieve this goal is to use the trilinear commutation relations, as it is formulated in the parastatistics, see Section 5.1, and was applied in consequent sections.

The direct calculation leads to the following trilinear commutation relations:

$$\left[\left[b_k^+, b_{k'}\right]_-, b_{k''}\right]_- = -2\delta_{kk'}\delta_{kk''}b_k, \tag{5.71}$$

$$\left[\left[b_k^+, b_{k'}\right]_-, b_{k''}^+\right]_- = 2\delta_{kk'}\delta_{kk''}b_k^+, \tag{5.72}$$

In the case $k = k' = k''$, these relations coincide with the trilinear commutation relations (5.1) and (5.1a) for the parafermi statistics. For different k, k', and k'' the relations are different. In the parafermi statistics in relation (5.1), corresponding to Eq. (5.71), instead of the two presented Kronecker symbols there is one; the similar difference is between Eqs. (5.1a) and (5.72). Thus, the pairon operators satisfy some modified parafermi statistics. According to Eq. (5.70) its rank $p = 1$, while the parastatistics of rank p satisfy Eqs. (5.10) and (5.11).

As follows from the definition of the particle number operator $\hat{N}_k$,

$$\hat{N}_k|N_k\rangle = N_k|N_k\rangle. \tag{5.73}$$

For the boson and fermion number operators the well-known expression $\hat{N} = a_k^+ a_k$ is valid. But it is not evident that the same expression is valid for the pairon number operator, cf. previous sections. In the parafermi statistics, the particle number operator is defined as

$$\hat{N}_k = \frac{1}{2}\left(\left[b_k^+, b_k\right]_- + p\right), \tag{5.74}$$

for pairons $p = 1$ and

$$\hat{N}_k = \frac{1}{2}\left(\left[b_k^+, b_k\right]_- + 1\right). \tag{5.75}$$

From the trilinear commutation relations (5.71) and (5.72), it follows that the operator (5.75) satisfies the commutation relations for the particle number operator, which were established earlier for bosons and fermions, see Eqs. (E.15) and (E.16), in Appendix E,

$$\left[\hat{N}_k, b_k^+\right]_- = b_k^+, \tag{5.76}$$

$$\left[\hat{N}_k, b_k\right]_- = -b_k. \tag{5.77}$$

For fermions Eq. (5.75) is equivalent to the standard expression $\hat{n}_k = c_k^+ c_k$; compare the derivation in Section 5.1 for parabose and parafermi statistics, which at $p = 1$ are reduced to the Bose (Fermi) statistics. Let us study it in the pairon case.

Using Eq. (5.69), the expression for the pairon number operator (5.75) can be written as

$$\hat{N}_k = \frac{1}{2}\left(\hat{n}_{k\alpha} + \hat{n}_{-k\beta}\right), \tag{5.78}$$

where $\hat{n}_{k\alpha} = c_{k\sigma}^+ c_{k\sigma}$ is the electron number operator. It is quite natural that the number of Cooper's pairs is two times less than the number of electrons forming pairs. It can be proved that from Eq. (5.78) it follows that the expression $\hat{N}_k = b_k^+ b_k$ for the pairon number operators is also valid. Let us do it.

The product $b_k^+ b_k$ is equal to the product of the fermion number operators $\hat{n}_{k\alpha}\hat{n}_{-k\beta}$, but in the general case

$$\hat{n}_{k\alpha}\hat{n}_{-k\beta} \neq \frac{1}{2}\left(\hat{n}_{k\alpha} + \hat{n}_{-k\beta}\right). \tag{5.79}$$

On the other hand, $n_{k\alpha}$ and $n_{-k\beta}$ are equal to 0 or 1; in this case, and only in this case, the left-hand part of Eq. (5.79) is equal to its right-hand part. Thus,

$$\frac{1}{2}\left(\hat{n}_{k\alpha} + \hat{n}_{-k\beta}\right) = b_k^+ b_k \tag{5.80}$$

and from Eqs. (5.78) and (5.80) it follows that although Cooper's pairs are neither bosons nor fermions, for the operators of their number, the traditional form $\hat{N}_k = b_k^+ b_k$ can be used.

If one substitutes the equality

$$\hat{n}_{k\alpha} + \hat{n}_{-k\beta} = 2b_k^+ b_k, \tag{5.81}$$

into the commutation relation (5.69), it transforms into

$$\left[b_k, b_k^+\right]_- = \left(1 - 2b_k^+ b_k\right) \tag{5.82}$$

or

$$\left[b_k, b_k^+\right]_+ = 1. \tag{5.83}$$

Thus, for equal **k**, the pairon operators obey the fermion commutation relations, while for different **k**, they obey the boson commutation relations. Despite the fact that each Cooper's pair has the total spin $S = 0$, the pairons are not bosons, because for equal momenta **k** they behave as fermions. However, for different **k**, the pairons obey the Bose–Einstein statistics and can occupy one energy level, that is, they can undergo the phenomenon of the Bose–Einstein condensation. However, in this case all electrons composed into the condensed Cooper pairs must have different momenta **k**.

Eq. (5.67) can be written similar to Eq. (5.82)

$$\left[b_k, b_{k'}^+\right]_- = \delta_{kk'}\left(1 - 2b_k^+ b_k\right). \tag{5.84}$$

The application of pairon operators to the vacuum state follows from their definition, Eq. (5.66),

$$b_k|0\rangle = 0, \quad b_k b_{k'}^+|0\rangle = \delta_{kk'}|0\rangle. \tag{5.85}$$

The relations (5.84) and (5.85) are sufficient for performing calculations using only the pairon operators. This allows to study problems in which the interactions between Cooper's pairs are also included, see Ref. [31].

5.4 Fractional Statistics

5.4.1 Eigenvalues of Angular Momentum in the Three- and Two-Dimensional Space

In standard textbooks on quantum mechanics the theory of angular momentum is usually discussed only in the three-dimensional (3D) space, using naturally the 3D rotation group, $\mathbf{R}_3$. The specific features of rotations in the two-dimensional (2D) space have not been discussed at all. The group $\mathbf{R}_3$ is a continuous group, in contrary to the discrete groups, as the permutation group or most of the point symmetry groups. From the theory of continuous groups (see Appendix C) it follows that any finite transformation of a continuous group can be represented as a succession of infinitely small transformations and a finite transformation is uniquely determined by specifying the infinitely small transformations. The latter are characterized by so-called *infinitesimal operators* of the group, I_ρ. Their number is equal to the number of group parameters.

In quantum mechanics the change in a wave function due to infinitely small rotations is expressed in terms of the angular momentum operator **J** [32]. The operator

J is a vector quantity and in the 3D space has, naturally, three independent components $\hat{J}_x$, $\hat{J}_y$, and $\hat{J}_z$. These components are related to the infinitesimal operators $\hat{I}_\rho$ by Eq. (C.24), see Appendix C,

$$\hat{J}_\rho = -i\hat{I}_\rho \tag{5.86}$$

and satisfy the commutation relations following from the commutation relations (C.22) for the infinitesimal operators $\hat{I}_\rho$. These commutation relations are represented in all textbooks on quantum mechanics. Namely,

$$\left[\hat{J}_x, \hat{J}_y\right] = i\hat{J}_z, \quad \left[\hat{J}_y, \hat{J}_z\right] = i\hat{J}_x, \quad \left[\hat{J}_z, \hat{J}_x\right] = i\hat{J}_y. \tag{5.87}$$

The operator $\mathbf{J}^2 = J_x^2 + J_y^2 + J_z^2$ is commute with the operators $\hat{J}_\rho$. This means that operators $\mathbf{J}^2$ and $\hat{J}_\rho$ (only one of its projection, let it be J_z) can possess a common set of eigenfunctions. Denote them as ψ_{jm} or in Dirac's notations as $|jm\rangle$.

As it is proved in quantum mechanics, see [32], from the commutation relations (5.87) the following equations are valid:

$$\hat{J}^2 |jm\rangle = j(j+1)|jm\rangle, \tag{5.88}$$

$$\hat{J}_z |jm\rangle = m|jm\rangle. \tag{5.89}$$

From (5.88) it follows that eigenvalues j of angular momentum can be only positive or zero. Since both directions of z-axis are equivalent, for each positive value of m exists the equal negative value $-m$. The projection of the angular momentum cannot be larger than the angular momentum j; therefore, the consequence of m is limited by the maximum value equals j and lays in the interval:

$$-j \le m \le j. \tag{5.90}$$

Basing on these properties and introducing the operators $\hat{J}_+$ and $\hat{J}_-$ that increase or decrease, respectively, the value m on unity, it can be proved [2, 32] that the length $2j$ of the interval (5.90) can take only integer values, which leaves for j (including the spin angular moment s) two possibilities: j is an integer or j is a half-integer. These two possibilities for spin, corresponding to the Bose–Einstein or Fermi–Dirac statistics, were accepted from the early years of quantum mechanics and presented in all textbooks on quantum mechanics. Surprisingly, but it takes more than 50 years to recognize that this is valid only in the 3D space or in a space with larger dimension.

As was demonstrated in 1976 by Leinaas and Myrheim [33], in the 2D space a continuum of intermediate cases between the boson and fermion cases can exist.

These intermediate cases indicate that in the 2D space the limitation on the spin by only integer or half-integer values stopped to be valid. The reason for the latter can be explained quite simply in the frame of the group theory formalism. The point is that the limitation on the angular momentum values comes from the commutation relations (5.87) for their operators. In the 2D space the rotation group, $\mathbf{R}_2$, is characterized by only one infinitesimal operator or related to it an angular momentum, which is perpendicular to the 2D space. Certainly the unique angular momentum operator commutes with itself and no limitations on its eigenvalues are arising. Between the rotational groups $\mathbf{R}_3$ and $\mathbf{R}_2$ there is a very essential distinction: the group $\mathbf{R}_3$ is non-Abelian and its representations are labeled by a discrete index, integer and half-integer values of the angular momentum, while $\mathbf{R}_2$ is an Abelian group and has a continuum of representations, corresponding to a continuum of allowed values of angular momentum. This is related not only to the total angular momentum, but to spin angular momentum as well. In the 2D space spin can take any rational value.

However, as we will see from the discussion below, a topological approach is more appropriate for studying consequences of the space dimension on the physical properties of identical particle systems. In the 2D space the results of these studies revealed rather complex, although very interesting, new phenomena. Leinaas and Myrheim [33] came to their conclusions not from group theory considerations, but using the topological approach. It should be mentioned that earlier in 1971 Laidlaw and DeWitt [34] applied the Feynman path integral formalism [35] to systems of identical particles. In this formalism the exchange of identical particle has a physical meaning as a continuous process in which each particle moves along a continuous path. This path dependence relates the exchange to the topology of the configuration space. Unfortunately, Laidlaw and DeWitt confined their attention to only three dimensions. Let us note that even earlier, the Feynman path formalism in connection with spin was applied by Schulman [36]. As mentioned the authors [34], their paper was induced by Schulman's publication [36].

Leinaas and Myrheim [33] considered the geometrical interpretation of the wave functions and were the first to study the influence of the space dimension on possible statistics. They took into account that in the two dimensions, the space is multiply connected [37], since in the 2D space there are an infinite number of exchange paths. In particular, they showed that for the two-particle system a continuously variable parameter ξ exists, which one can choose to be the phase angle:

$$P=\exp(i\xi), \tag{5.91}$$

where the parameter ξ is a real number, independent of the point on the 2D surface. For boson systems $\xi=0$ and for fermion systems $\xi=\pi$, while for values of ξ in the interval

$$0<\xi<\pi \tag{5.92}$$

there are an infinite number of intermediate cases. As an illustration, the authors [33] solved the spectrum of two particles in the 2D harmonic oscillator potential and showed that as ξ changes from 0 to π, there is a continuous transition from the boson to fermion spectrum. Whereas for a two-particle system in the 3D space, which is simply connected, they obtained only two permitted types of particles: bosons ($\xi = 0$) and fermions ($\xi = \pi$).

For N-particle system, Leinaas and Myrheim [33] obtained in the 3D space also only two permitted statistics: the Bose–Einstein and Fermi–Dirac statistics, as in the two-particle case. In this connection I would like to stress that this part of the study [33] is wrong. It is based on the paper by Mirman [38]. In that publication Mirman made an incorrect statement that the exchange of identical particles does not have an effect on the configuration of one-particle wave functions, since it leads to the same configuration, and the widely used definition of the indistinguishability principle in the form (2.5) has no physical sense. However, it is easy to show that the considerations of Mirman [38] are unphysical. From his ideas, it follows that we cannot use in physics the conception of the symmetry group of the Hamiltonian; not only the permutation group, but also the point symmetry groups, since the application to a molecule of some of its symmetry transformations bring the molecule into itself. Thus, according to Mirman's statement, the point group symmetry operations have also no sense.

Unfortunately, the wrong ideas presented in Ref. [38] were completely accepted first by Leinaas and Myrheim [33] and then in many consequent studies in the fractional statistics field, see, for instance, Refs. [39, 40]. So in section 2.3 in the book [40], Khare, citing Mirman [38], writes: "If the particles are *strictly identical*, the word permutation has no physical meaning, since a given configuration and the one obtained by the permutation of the particle coordinates are merely two different ways of describing the *same* particle configuration." In this connection it is reasonable to ask: to what kind of particles, if not to strictly identical ones, the permutation group can be applied?!

As a result, during a long time in the fractional statistics community it was accepted that for any system of identical particle the only one configuration of one-particle wave functions with fixed quantum numbers can be constructed; the permutation degeneracy is absent. From this it was concluded that in the case of identical particles, the Hilbert space of the N-particle system is *always one-dimensional* (1D). Certainly, in this Hilbert space only the 1D representations exist, that is, in the 3D space only the symmetric and antisymmetric wave functions are allowed. However, in Section 3.1 we demonstrated that the similar proof, represented by Eqs. (3.2)–(3.4), is incorrect from the mathematical viewpoint and from physical as well. In general case, the phase angle in Eq. (5.91) cannot be a simple real number. In the 3D space, the angle is a function of spatial coordinates and permutation, see Eq. (3.9).

Let us stress that in the 2D space the presentation of the phase factor for N-particle system in the form (5.91) with ξ as a real number, as it was accepted for the two-anyon system, also may not be valid in all cases. The movement of each anyon depends, in general, upon the location of *all* others, thus it can be expected to have a more complicated behavior. We will return to this matter in the next subsection.

5.4.2 Anyons and Fractional Statistics

In this section I discuss the concepts of anyons and fractional statistics stressing only the most important features from my viewpoint. One can find a more detailed information in the book by Khare [40].

The new results revealed by Leinaas and Myrheim [33] do not attract an attention in physical community until the publications by Wilczek [41, 42]. Wilczek considered particles with the electric charge q orbiting around a tube (solenoid) with a magnetic flux Φ. When a flux Φ runs through the solenoid, the integer[1] orbital angular momentum l_z is changed on $\Delta l_z = -q\Phi/2\pi$. Thus, the angular momentum in considered system depends on the magnitude of the flux Φ,

$$l_z = \text{integer} - q\Phi/2\pi. \tag{5.93}$$

The interchange of the flux-tube-particle composites leads, as it must be in all gauge-invariant theories, to a phase factor $\exp(iq\Phi)$. If the value of l_z given by Eq. (5.93) is an integer or a half-integer, the statistics is normal. But in intermediate cases, the composites cannot be described as fermions or as bosons. They have unusual statistics interpolating bosons and fermions. Since interchange of two of these quasiparticles can give *any* phase, Wilczek called them as *anyons*.

This name has since generalized to any 2D objects with unusual statistics. The latter has been called *fractional statistics*. The name of statistics is connected with the phase factor (5.91). If for $\xi = 0$ (π), the system obeys the standard boson (fermion) statistics, in the intermediate cases (5.92) ξ is a *fractional* multiple of π

$$\frac{\xi}{\pi} = \frac{1}{m}, \tag{5.94}$$

where m is an integer number.

Thus, the theoretical possibility of existence of anyons and fractional statistics does not contradict quantum mechanics. Then it is important to know if they really exist in our nature? The answer to this question is positive. At least one physical

[1] As we discussed above, in the 3D space the total angular momentum, which includes the orbital and spin angular momenta, can take integer and half-integer values. However, the orbital angular momentum can take only integer values [32].

phenomenon has been explained using these exotic conceptions. It is the fractional quantum Hall effect (FQHE) [43].

As it was observed by Tsui et al. [44], after the condensation of 2D electron gas in a GaAs-$Ga_xAl_{1-x}As$ hetorostructure at low temperature and large magnetic field, the Landau level was 1/3 filled. This experiment was explained by Laughlin [45] as the FQHE, in which an incompressible quantum fluid of the elementary excitations is formed. This effect is controlled by the occupation ν of the lowest Landau level; the occupation number is a fraction, which can be presented as

$$\nu = \frac{p}{q}, \tag{5.95}$$

where p and q are integers, and q being odd. The quasiholes and quasiparticles in the quantum fluid are fractionally charged with charges $e^* = \pm e/q$.

In his theory Laughlin [45] did not discuss the statistics of his fractionally charged excitations. It was done in the study by Arovas et al. [46]. They rigorously proved that the elementary excitations (quasiparticles) in FQHE are obeyed the fractional statistics and this is related with their fractional electric charge. At the same time the general study of the fractional statistics and its connection with FQHE was performed by Haldane [47] and Halperin [48].

In 1988, Laughlin [49, 50] suggested another candidate for the realization of fractional statistics and anyons in our nature. He discussed the possibility of the anyon mechanism of high-T_c superconductivity in cuprates, in CuO_2 planes. This suggestion attracted a great interest among the physical community [51–54]. The main discussion was connected with the symmetry properties of the ground superconducting state if the mechanism of superconductivity has an anyonic nature. In the first responses [51, 52], it was pointed out that a striking property of this anyon superconductor is that its ground state is characterized by the violation of parity (P) and time-reversal (T) symmetries. In FQHE materials the time reversal symmetry is naturally broken by the magnetic field, while superconductors are invariant in respect to T symmetry. As was stressed by March-Russell and Wilczek [52], if the anyon mechanism is realized in the high-T_c superconductors, it can be tested in experiment. Such test was performed by two groups [55, 56] using different experimental methods. In both experiments the obtained results were negative, the broken time reversal symmetry was not observed. Thus the FQHE structures stay till today the unique materials where the anyons and fractional statistics are realized. It is important to note that the predicted in the FQHE theory quasiparticles (excitations) with fractional charges were detected by three different experimental methods [57–59].

The collective properties of N-anyon systems were studied by different authors. Already in the first study by Leinaas and Myrheim [33], the authors solved the spectrum of two anyons in the 2D harmonic oscillator potential and, particularly, showed that it is not related to the corresponding single anyon spectrum. In 1984,

Wu [60] obtained the exact solution for three anyons in the 2D harmonic oscillator potential. In contrary to the two-anyon case where a continuous interpolation between the bosonic and fermionic spectra was obtained [33], the exact three-anyon state, which Wu [60] obtained starting from the three-boson ground state, does not interpolate to the three-fermion ground state. The energy obtained at $\xi = \pi$ exceeds the energy of the three-fermion ground state that manifests the level crossing with some excited state. As Khare [40] correctly noted in this connection: "This gave a first hint that the multi-anyon problems are going to be highly nontrivial." Subsequent studies completely confirmed this notice. It became clear that the anyons cannot be considered as independent objects. Even in the absence of dynamical interactions in any anyon system some "statistical" interaction always presents and the ideal gas of anyons cannot exist in principle.

As we discussed in the preceding subsection, in the 2D space it is quite convenient to use the Feynman path integral formalism, which relates the exchange of particles (or quasiparticles) and following from it possible statistics to the topology of the configuration space. In this formalism any exchange is considered as a continuous process, in which both particles are moving along its trajectories, thus the time should be added to two spatial coordinates, which is denoted as (2 + 1)D. Two exchanges are not topologically equivalent to the zero exchange, as it is in the 3D space. In the path-integral language it means that if two particles are interchanged twice in a clockwise manner, their trajectory involves a nontrivial winding, and the system does not necessarily come back to the same state. If in the 3D space exchanges are just permutations and all exchange paths are homotopic to one another, in the 2D space exchanges involve braiding, and one must distinguish topologically inequivalent trajectories leading to the same permutations of the particles. Hence, in the 2D space the permutation group cannot be applied; it is replaced by the so-called *braid group* [61–63].

The topological classes of trajectories, which take the particles from initial positions $\boldsymbol{r}_1, \boldsymbol{r}_2, \ldots, \boldsymbol{r}_N$ at time t_1 to final positions $\boldsymbol{r}_1', \boldsymbol{r}_2', \ldots, \boldsymbol{r}_N'$ at time t_2, are in one-to-one correspondence with the elements of the braid group $\boldsymbol{B}_N$; a vector $\boldsymbol{r}_i$ is the 2D vector defining the position of particle i. An element of the braid group can be visualized by representing the particle trajectories as world lines (or *strands*) originating at initial positions and terminating at final positions, with the vertical time direction, as shown in Fig. 5.2 for three particles. When drawing the trajectories, one must be careful to distinguish when one strand passes over or under another, corresponding to a clockwise or counterclockwise exchange.

Let us denote by σ_i the braid operation corresponding to the counterclockwise exchange of neighboring particles i and $i+1$. Then σ_i^{-1} is a clockwise exchange of particles i and $i+1$. For operations σ_i the following relations are valid [63]:

$$\sigma_i \sigma_j = \sigma_j \sigma_i \quad \text{for} \quad |i-j| \geq 2, \tag{5.96}$$

$$\sigma_i \sigma_{i+1} \sigma_i = \sigma_{i+1} \sigma_i \sigma_{i+1} \quad \text{for} \quad 1 \leq i \leq n-1. \tag{5.97}$$

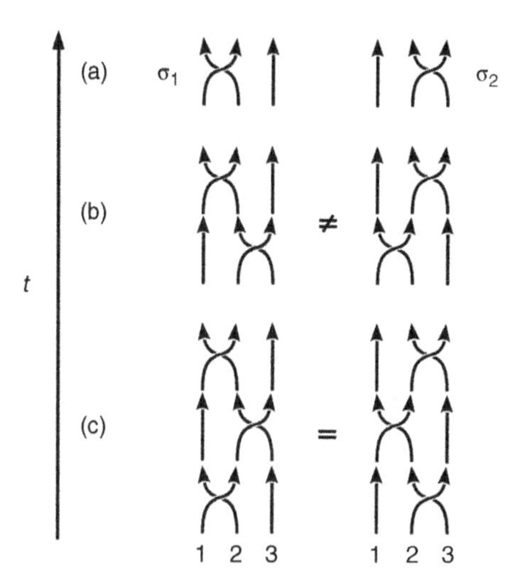

Fig. 5.2 Graphical representation of elements of the braid group. (a) The two elementary braid operations σ_1 and σ_2. (b) Evidence that $\sigma_2\sigma_1 \neq \sigma_1\sigma_2$; hence the braid group is non-Abelian. (c) Graphical representation of Eq. (5.97)

Eq. (5.96) is evident, since the elements σ_i and σ_j are acting on different particles, because they exchange only two neighboring particles. The validness of the braid relation (5.97) follows from Fig. 5.2c.

The principal difference between the permutation group π_N and the braid group $\boldsymbol{B}_\mathrm{N}$ is that in the latter $\sigma^2 \neq 1$. As a result, the permutation group is finite, while the braid group is infinite, even for two particles. Another important property of the braid group follows from Fig. 5.2b:

$$\sigma_2\sigma_1 \neq \sigma_1\sigma_2. \tag{5.98}$$

The braid group is *non-Abelian*, that is, it can have multidimensional irreducible representations; a more detailed information can be found in review [63].

As we discussed above, in early studies of identical anyon systems it was accepted that anyons are always described by the one-dimensional, that is, Abelian, representations, no other possibilities were considered. In this case, if one anyon is exchanged with the other, the wave function for these Abelian anyons is multiplied on the nontrivial phase factor,

$$\psi(r_2, r_1) = e^{i\xi}\psi(r_1, r_2), \tag{5.99}$$

which is not necessarily equal to ±1. Each real ξ in the interval (5.92) can correspond to some fractional statistics.

In the first studies by Wilczek [41, 42], only two anyons were considered with nontrivial Abelian representations (5.99). However, in the general case of N-anyon system, the anyons can be also be described by the *multidimensional representations* of the braid group, see above. Let it be an f-dimensional representation. Then these non-Abelian anyons are described by a set of f functions, which form a basis for that representation. Under the action of the group elements, the functions transform only into each other according to Eq. (A.11), Appendix A. Thus, instead of the phase factors (5.99), the wave functions of non-Abelian anyons transform via a square matrix of order f. These matrices form an f-dimensional representation of the braid group. Certainly, for the realization of multidimensional representations, the Hilbert space for anyons must not be one-dimensional.

First the possibility of non-Abelian anyons was discussed by Moore and Read [64] in connection with FQHE. They suggested that the states with $\nu = 1/2, 1/4, \ldots$ obey non-Abelian fractional statistics. They pointed out that these states can be interpreted as the BCS pair states of composite fermions (see also Ref. [65]), which can be of two types: the spin-singlet d-waves and a p-wave polarized state; the latter they called the Pfaffian state.

In the early theories of FQHE by Laughlin [45] and then by Tao [66] it was formulated by the so-called "odd-denominator rule" for the occupation numbers ν of the Landau levels, see Eq. (5.95). Just this type of fractional charges $e^* = \pm e/q$ with $q = 3$, 5, 7, and 9 was detected in experiments [57–59]. Meanwhile, already in 1987 it was detected a quantized Hall plateau at $\nu = 5/2$ [67]. The state with $\nu = 5/2$ has been discussed as a most probable candidate for the non-Abilean quantum Hall state; see review by Read [68]. However, without experimental confirmation it has been stayed only as a theoretical possibility. Several experimental verifications were proposed and discussed in detail [69–72]. In experimental studies of the 5/2 fractional quantum Hall (FQH) state the quasiparticles with the charge $e/4$ were observed [73–75]. Unfortunately, the definite conclusion about the non-Abelian nature of the measured $\nu = 5/2$ state still could not be done; see discussion in Ref. [76].

Last years, the FQHE became an object of intense experimental and theoretical studies. So recently the FQH states were revealed in graphene [77–79], where many new physical properties were detected [80]. Nevertheless, to the best of my knowledge, the existence of non-Abelian anyons is still an open question.

It is important to stress that the possibility of the fractional statistics in the 2D space is related only to quasiparticles (excitation states of the particle systems). If electrons are confined in the plane they remain fermions and obey the Fermi–Dirac statistics. Thus, only the systems of quasiparticles (quasiholes) can obey the fractional statistics. The immanent property of quasiparticle systems, in particular the anyon system, is that they in principle may not form an ideal (noninteracting) system. The motion of one anyon depends, in general, on the location of others. Anyon systems even in the absence of dynamical interactions are characterized by some statistical interaction. This immanent property is similar to the property of the quasiparticles in a periodical lattice obeying the modified parafermi statistics, see Section 5.2. Although for anyons the physical situation is even more complex that is reflecting in the hard problems arising in the calculations of N-anyon systems.

One more comment. In contrary to the quasiparticles obeying the modified parastatistics, which are widely realized in Nature as collective excitations: excitons and magnons [9], holes, and other collective excitations in solids, see Section 5.2, the excitations obeying the fractional statistics are realized only in the FQHE. However, in the physico-mathematical community there are great expectations that the possible non-Abelian FQH states will be applied in future in the topological quantum computation, first proposed by Kitaev [81] and described in detail in review [63].

References

[1] H.S. Green, *Phys. Rev.* **90**, 270 (1953).

[2] A.S. Davydov, *Theory of Molecular Excitons*, McGraw-Hill, New York, 1962.

[3] S. Kamefuchi and Y. Takahachi, *Nucl. Phys.* **36**, 177 (1962).

[4] C. Rayan and E.C.G. Sudarshan, *Nucl. Phys.* **47**, 207 (1963).

[5] Y. Ohnuki and S. Kamefuchi, *Quantum Field Theory and Parastatistics*, Springer-Verlag, Berlin, 1982.

[6] N.A. Chernikov, *Acta Physica Polon.* **21**, 52 (1962).

[7] O.W. Greenberg, *Phys. Rev. Lett.* **13**, 598 (1964).

[8] A.B. Govorkov, *Sov. J. Part. Nucl.* **14**, 520 (1983).

[9] I.G. Kaplan, *Theor. Math. Phys.* **27**, 254 (1976).

[10] A.N. Avdyugin, Yu.D Zavorotnev, and L.N. Ovander, *Sov. Phys. Solid State* **25**, 1437 (1983).

[11] B.A. Nguyen, *J. Phys. C Solid State Phys.* **21**, L1209 (1988).

[12] D.I. Pushkarov, *Phys. Status Solidi b* **133**, 525 (1986).

[13] A. Nguyen and N.C. Hoang, *J. Phys. Condens. Matter* **2**, 4127 (1990).

[14] I.G. Kaplan and O. Navarro, *J. Phys. Condens. Matter* **11**, 6187 (1999).

[15] I.G. Kaplan and O. Navarro, *Physica C* **341–348**, 217 (2000).

[16] I.G. Kaplan, J. Soullard, J. Hernandez-Cobos, and R. Pandey, *J. Phys. Condens. Matter* **11**, 1049 (1999).

[17] E. Dagotto, *Rev. Mod. Phys.* **66**, 763 (1994).

[18] I.G. Kaplan and V.G. Plotnikov, *Khim. Vysok. Energii* **1**, 507 (1967).

[19] I.G. Kaplan, *Intermolecular Interactions: Physical Picture, Computational Methods and Model Potentrials*, John Wiley & Sons, Ltd, Chichester, 2006.

[20] J.E. Hirsch, *Phys. Rev. B* **48**, 3327 (1993).

[21] S.V. Tyablikov, *Methods of Quantum Theory of Magnetism*, Plenum Press, New York, 1967.

[22] P.W. Anderson, *Science* **235**, 1196 (1987).

[23] S.A. Kivelson, D.S. Rokhsar, and J.P. Sethna, *Phys. Rev. B* **35**, 8865 (1987).

[24] P. W. Anderson, G. Baskaran, Z. Zou, and T. Hsu, *Phys. Rev. Lett.* **58**, 2790 (1987).

[25] F. Dyson, *Phys. Rev.* **102**, 1217 (1956).

[26] S.T. Belyaev, *Sov. Phys.—JETP* **7**, 289 (1958).

[27] I.G. Kaplan and M.A. Ruvinsky, *Sov. Phys.—JETP* **44**, 1127 (1976).

[28] M.W. Schafer et al., *Phys. Rev. B* **39**, 2914 (1989).

[29] H. Zhang and H. Sato, *Phys. Rev. Lett.* **70**, 1697 (1993).

[30] J.R. Schrieffer, *Theory of Superconductivity*, Addison-Wesley, Redwood City, California, 1988.

[31] I.G. Kaplan, O. Navarro, and J.A. Sanchez, *Physica C* **419**, 13 (2005).

[32] L.D. Landau and E.M. Lifshitz, *Quantum Mechanics (Nonrelativistic Theory)*, 3rd edn., Pergamon Press, Oxford, 1977.

[33] J.M. Leinaas and J. Myrheim, *Nuovo Cimento* **37B**, 1 (1977).
[34] M.G.G. Laidlaw and C.M. DeWitt, *Phys. Rev. D* **3**, 1375 (1971).
[35] R.P. Feynman and A.R. Hibbs, *Quantum Mechanics and Path Integrals*, McGraw-Hill, New York, 1965.
[36] L. Schulman, *Phys. Rev.* **176**, 1558 (1968).
[37] I.M. Singer and J.A. Thorpe, *Lecture Notes on Elementary Topology and Geometry*, Scott, Foresman, New York, 1967.
[38] R. Mirman, *Nuovo Cimento* **18B**, 110 (1973).
[39] G.S. Canright and S.M. Girvin, *Science* **247**, 1197 (1990).
[40] A. Khare, *Fractional Statistics and Quantum Theory*, 2nd edn., World Scientific, Singapore, 2005.
[41] F. Wilczek, *Phys. Rev. Lett.* **48**, 1144 (1982).
[42] F. Wilczek, *Phys. Rev. Lett.* **49**, 957 (1982).
[43] R.E. Prange and S.M. Girvin, Eds., *The Quantum Hall Effect*, 2nd edn., Springer, New York, 1990.
[44] D.C. Tsui, H.L. Stormer, and A.C. Gossard, *Phys. Rev. Lett.* **48**, 1559 (1982).
[45] R.B. Laughlin, *Phys. Rev. Lett.* **50**, 1395 (1983).
[46] D. Arovas, J.R. Schrieffer, and F. Wilczek, *Phys. Rev. Lett.* **53**, 722 (1984).
[47] F.D.M. Haldane, *Phys. Rev. Lett.* **51**, 605 (1983); 67, 937 (1991).
[48] B.I. Halperin, *Phys. Rev. Lett.* **52**, 1583 (1984).
[49] R.B. Laughlin, *Phys. Rev. Lett.* **60**, 2677 (1988).
[50] R.B. Laughlin, *Science* **242**, 525 (1988).
[51] A. Kivelson and D.S. Rokhsar, *Phys. Rev. Lett.* **61**, 2630 (1988).
[52] J. March-Russell and F. Wilczek, *Phys. Rev. Lett.* **61**, 2066 (1988).
[53] X.G. Wen and A. Zee, *Phys. Rev. Lett.* **62**, 2873 (1989).
[54] Y.H. Chen, F. Wilczek, E. Witten, and B.I. Halperin, *Int. J. Mod. Phys.* **3**, 1001 (1989).
[55] R.F. Kiefl et al., *Phys. Rev. Lett.* **64**, 2082 (1990).
[56] S. Spielman et al., *Phys. Rev. Lett.* **65**, 123 (1990).
[57] R.G. Clark et al., *Phys. Rev. Lett.* **60**, 1747 (1988).
[58] J.A. Simmons et al., *Phys. Rev. Lett.* **63**, 1731 (1989).
[59] A.M. Chang and J.E. Cummingham, *Solid State Commun.* **72**, 651 (1989).
[60] Y.-S. Wu, *Phys. Rev. Lett.* **53**, 111 (1984).
[61] S. Moram, *The Mathematical Theory of Knots and Braids*, North-Holland, Amsterdam, 1983.
[62] Y.-S. Wu, *Phys. Rev. Lett.* **52**, 2103 (1984).
[63] C. Nayak, S.H. Simon, A. Stern, M. Freedman, and S. Das Sarma, *Rev. Mod. Phys.* **80**, 1083 (2008).
[64] G. Moore and N. Read, *Nucl. Phys. B* **360**, 362 (1991).
[65] F.D.M. Haldane and E.H. Rezayi, *Phys. Rev. Lett. E* **60**, 956, 1886 (1988).
[66] R. Tao, *Phys. Rev. B* **23**, 2937 (1986).
[67] R.L. Willett et al., *Phys. Rev. Lett.* **59**, 1779 (1987).
[68] N. Read, *Physica B* **298**, 121 (2001).
[69] S. Das Sharma, M. Freedman, and C. Nayak, *Phys. Rev. Lett.* **94**, 166802 (2005).
[70] A. Stern and B.I. Halperin, *Phys. Rev. Lett.* **96**, 016802 (2006).
[71] P. Bonderson, A. Kitaev, and K. Shtengel, *Phys. Rev. Lett.* **96**, 016803 (2006).
[72] X. Wen, Z. Hu, E.H. Rezayi, and K. Yang, *Phys. Rev. B* **77**, 165316 (2008).
[73] M. Dolev, M. Heiblum, V. Umansky, A. Stern, and D. Mahalu, *Nature* **452**, 829 (2008).
[74] I. Radu et al. *Science* **320**, 899 (2008).
[75] R.L. Willett, L.N. Pfeiffer, and K.W. West, *Proc. Natl. Acad. Sci. U. S. A.* **106**, 8853 (2009).
[76] B. Stern, B. Rosenow, R. Ilan, and B.I. Halperin, *Phys. Rev. B* **82**, 085321 (2010).
[77] X. Du, I. Skachko, F. Duerr, A. Luican, and E. Andrei, *Nature* **462**, 192 (2009).
[78] K.I. Bolotin, F. Ghahari, M.D. Shulman, H.L. Stormer, and P. Kim, *Nature* **462**, 196 (2009).
[79] B.E. Feldman, B. Krauss, J.H. Smet, and A. Yacoby, *Science* **337**, 1196 (2012).
[80] D.A. Abanin, B.E. Feldman, A. Yacoby, and B.I. Halperin, *Phys. Rev. B* **88**, 115407 (2013).
[81] A.Y. Kitaev, *Ann. Phys. (Berlin)* **303**, 2 (2003).

Appendix A

Necessary Basic Concepts and Theorems of Group Theory[1]

A.1 Properties of Group Operations

A.1.1 Group Postulates

A set of elements $A, B, \ldots$, is said to form a group **G** if it satisfies the following four conditions:

1. A "multiplication law" for the elements is specified, that is, a rule is prescribed according to which each pair of elements P, Q is placed into correspondence with some element R which is also contained in the same set. Element R is termed the product of the elements P and Q, and is written in the form

$$R = PQ. \tag{A.1}$$

[1] For a more detailed account of the group theory see in my book [1] or in many others, for example, in books [2–4].

The Pauli Exclusion Principle: Origin, Verifications, and Applications, First Edition. Ilya G. Kaplan.

2. The product of the factors is associative:

$$P(QR) = (PQ)R, \tag{A.2}$$

that is, in order to specify a product uniquely, it is sufficient to specify the order of the factors.
3. Among the elements of the group there is a unit element E possessing the property

$$EQ = QE = Q, \tag{A.3}$$

for any Q belonging to the group (the condition that the element Q belongs to the group **G** is denoted symbolically as follows: $Q \in \mathbf{G}$).
4. For every element $Q \in \mathbf{G}$ there exists an inverse element $Q^{-1} \in \mathbf{G}$ that satisfies the equation

$$Q^{-1}Q = QQ^{-1} = E. \tag{A.4}$$

The element which is the inverse of a product of elements is given by

$$(PQ)^{-1} = Q^{-1}P^{-1}, \tag{A.5}$$

as can easily be seen on multiplying PQ by $Q^{-1}P^{-1}$ and using the associative rule for products.

The product of elements in a group is generally noncommutative, that is,

$$PQ \neq QP.$$

If all elements of a group satisfy the equation $PQ = QP$, the group is said to be *Abelian*. The cyclic groups are a particular case of Abelian groups in which all the elements are obtained by successively raising the power of one of the elements, that is, the n elements of a cyclic group can be represented as follows:

$$A,\ A^2,\ A^3, \ldots,\ A^n \equiv E. \tag{A.6}$$

The following theorem is valid for the elements of a group:

If G_a *runs through all the elements of a group* **G** *and* G_0 *is some fixed element of* **G**, *then the product* G_0G_a *(or* G_aG_0*) also runs though all the elements of the group and, moreover, does so once only.*

Any element G_b of the group can, in fact, be obtained by multiplying G_0 from the right by $G_a = G_0^{-1}G_b$. Furthermore, the product G_0G_a cannot occur more than once since if $G_0 G_a = G_0 G_b$, then by multiplying this equation from the left by G_0^{-1} we would obtain $G_a = G_b$. Hence, for different G_a all $G_0 G_a$ are different.

From this theorem it follows that an arbitrary function of the group elements when summed over all the elements of the group is an invariant:

$$\sum_{G_a} f(G_a) = \sum_{G_a} f(G_0 G_a) = \sum_{G_b} f(G_b). \tag{A.7}$$

A.1.2 Examples of Groups

1. The set of all vectors in a three-dimensional space constitutes a group in respect to the operation of addition. In this case the operation of multiplying the elements of the group is the vector addition. The operation of vector addition possesses the associative property, and the unit element is represented by a vector of zero length. Mutually inverse elements of the group are represented by vectors equal in magnitude and opposite to one another in direction.
2. Another example of a group is formed by the set of spatial transformations of an equilateral triangle, under which the triangle is sent into itself. Such transformations are called *symmetry operations* (asymmetric bodies cannot be made to coincide with themselves under any such transformations apart from the identity). There are six distinct nonequivalent symmetry operations for an equilateral triangle and for these we choose the following (see Fig. A.1):
   E is the identity operation, which leaves the triangle unchanged.
   C_3 and C_3^2 are clockwise rotations by 120° and 240°, respectively, about an axis perpendicular to the plane of the triangle and passing through its center of gravity.

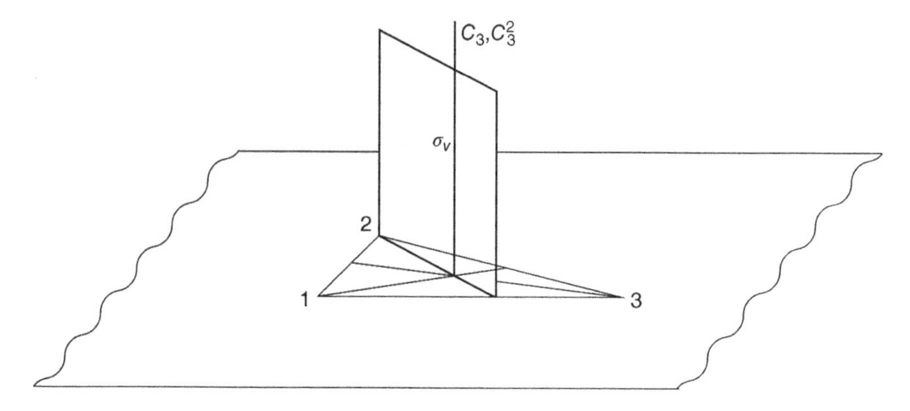

Fig. A.1 Explanations in text

Table A.1 Multiplication table for the group $\mathbf{C}_{3v}$

	E	C_3	C_3^2	$\sigma_v^{(1)}$	$\sigma_v^{(2)}$	$\sigma_v^{(3)}$
E	E	C_3	C_3^2	$\sigma_v^{(1)}$	$\sigma_v^{(2)}$	$\sigma_v^{(3)}$
C_3	C_3	C_3^2	E	$\sigma_v^{(2)}$	$\sigma_v^{(3)}$	$\sigma_v^{(1)}$
C_3^2	C_3^2	E	C_3	$\sigma_v^{(3)}$	$\sigma_v^{(1)}$	$\sigma_v^{(2)}$
$\sigma_v^{(1)}$	$\sigma_v^{(1)}$	$\sigma_v^{(3)}$	$\sigma_v^{(2)}$	E	C_3^2	C_3
$\sigma_v^{(2)}$	$\sigma_v^{(2)}$	$\sigma_v^{(1)}$	$\sigma_v^{(3)}$	C_3	E	C_3^2
$\sigma_v^{(3)}$	$\sigma_v^{(3)}$	$\sigma_v^{(2)}$	$\sigma_v^{(1)}$	C_3^2	C_3	E

$\sigma_v^{(1)}, \sigma_v^{(2)}, \sigma_v^{(3)}$ are reflections in planes perpendicular to the plane of the triangle and passing through its medians.

All other symmetry operations are equivalent to one of those just enumerated. Thus, for example, a rotation by 180° about an axis passing through a median is equivalent to a reflection σ_v.

Let us determine the products of all possible pairs of the six symmetry operations of the equilateral triangle given in Fig. A.1, and let us put the results in the form of a table (Table A.1). The operations which act on the triangle first are placed on the top line of the table. From this can be seen that the successive application of any two operations is equivalent to a single operation from the same set. For every operation Q there is an inverse operation Q^{-1}, which leads to the identify transformation E. By using the table one easily verifies that the associative rule is also obeyed. The symmetry operations of a triangle therefore form a group which is usually denoted by $\mathbf{C}_{3v}$. The group $\mathbf{C}_{3v}$ is one of the so-called *point groups*.

A.1.3 Isomorphism and Homomorphism

Groups are classified either as *finite* or *infinite* depending upon whether they contain a finite or an infinite number of elements. The number of elements in a group is termed its *order*. Two groups $\mathbf{G}$ and $\mathbf{G}'$ of the same order are termed *isomorphic* if a one-to-one correspondence between their elements can be defined such that if the element $A' \in \mathbf{G}'$ corresponds to $A \in \mathbf{G}$ and $B' \in \mathbf{G}'$ corresponds to $B \in \mathbf{G}$, then the element $C' = A'B'$ corresponds to $C = AB$. Isomorphic groups are identical as far as their abstract group properties are concerned, although the actual meaning of their respective elements may be quite different.

The groups $\mathbf{C}_{3v}$ introduced in the previous section and the permutation group π_3 (see Appendix B) form an example of isomorphous groups. We can regard each symmetry operation of the triangle as the corresponding permutation of its vertices, and so derive a one-to-one correspondence between the elements of the two groups:

$$\mathbf{C}_{3v}: \; E \quad C_3 \quad C_3{}^2 \quad \sigma_v^{(1)} \quad \sigma_v^{(2)} \quad \sigma_v^{(3)}$$

$$\pi_3 \quad : \; I \quad P_{132} \quad P_{123} \quad P_{23} \quad P_{13} \quad P_{12}$$

This is a particular case of a more general theorem (*Cayley's theorem*) which states:

any finite group of order n is isomorphous with a subgroup of a permutation group $\boldsymbol{\pi}_n$.

If for each element A of a group $\mathbf{G}$ there is a corresponding element A' of a group $\mathbf{G}'$ so that if $AB = C$, then $A'B' = C'$, the two groups are said to be *homomorphic*. In contrast to isomorphism, homomorphism does not require a one-to-one correspondence between the elements of the groups, and one element of the group $\mathbf{G}'$ may correspond to several elements of $\mathbf{G}$.

A.1.4 Subgroups and Cosets

If it is possible to select from a group $\mathbf{G}$ some set of elements which, with the same multiplication law, form a group $\mathbf{H}$ among themselves, then $\mathbf{H}$ is called a subgroup of $\mathbf{G}$. Every group possesses a trivial subgroup that consists of the one unit element of the group. Henceforth, when referring to a subgroup a nontrivial subgroup will always be implied.

Let the element G_1 belong to a finite group $\mathbf{G}$, but not to its subgroup $\mathbf{H}$, which consists of h elements. We multiply all the elements of $\mathbf{H}$ (e.g., from the left) by G_1 and obtain some set of h elements which we denote by $G_1\mathbf{H}$. Now, no element of $G_1\mathbf{H}$ belongs to $\mathbf{H}$. Otherwise we would have for some two elements $H_a, H_b \in \mathbf{H}$ the equation $G_1 H_a = H_b$, or $G_1 = H_b H_a^{-1}$, that is, $G_1 \in \mathbf{H}$, in contradiction with our initial assumption. We now take some element $G_2 \in \mathbf{G}$ not belonging to either $\mathbf{H}$ or $G_1\mathbf{H}$ and form the set $G_2\mathbf{H}$. In an analogous way one can show that $G_1\mathbf{H}$ and $G_2\mathbf{H}$ do not possess any elements in common. If $\mathbf{H}$, $G_1\mathbf{H}$, and $G_2\mathbf{H}$ do not exhaust all the elements of the group, we form $G_3\mathbf{H}, \ldots,$ until all the elements of $\mathbf{G}$ are divided into m sets:

$$\mathbf{H},\, G_1\mathbf{H},\, G_2\mathbf{H}, \ldots, G_{m-1}\mathbf{H}. \tag{A.8}$$

The order of the group is therefore given by $g = mh$. As a result, we arrive at a theorem known as *Lagrange's theorem*:

The order of a subgroup of a finite group is a divisor of the order of the group.

From this theorem it obviously follows that a group whose order is a prime number does not possess any subgroups.

The division of the group elements into the sets (A.8) is uniquely determined on specifying the subgroup **H** since any element of the set $G_k\mathbf{H}$ may play the role of G_k. Thus let $G_k' = G_kH_a$, where H_a is an arbitrary element of the subgroup **H**. The set of elements $G_k'\mathbf{H} \equiv G_kH_a\mathbf{H} = G_k\mathbf{H}$, the last equality following from the theorem of section 1.1.

The sets (A.8) are called the *left cosets* of the subgroup **H**. *Right cosets* are defined in an analogous manner. The number of cosets of a subgroup is called the index of the subgroup. With the exception of **H** none of the cosets $G_k\mathbf{H}$ forms a subgroup since they do not contain the unit element.

As an example, we consider the set of permutations I and P_{12} from the group $\boldsymbol{\pi}_3$. This set forms a subgroup which we denote by $\boldsymbol{\pi}_2$. The index of the subgroup m is equal to $g/h = 6/2 = 3$. There are therefore three cosets. The left cosets of the subgroup $\boldsymbol{\pi}_2$ are given by

$$\boldsymbol{\pi}_2 \quad : \quad I, P_{12};$$

$$P_{13}\boldsymbol{\pi}_2 : \quad P_{13}, P_{123};$$

$$P_{23}\boldsymbol{\pi}_2 : \quad P_{23}, P_{132}.$$

A.1.5 Conjugate Elements. Classes

Two elements A and B are said to be conjugate if $A = QBQ^{-1}$, where Q is an element of the same group. If two elements A and B are each conjugate to a third element C, then they are also conjugate to each other. Thus it follows from $A = QCQ^{-1}$ and $B = RCR^{-1}$ that $C = Q^{-1}AQ$, whence $B = RQ^{-1}AQR^{-1} = (RQ^{-1})A(RQ^{-1})^{-1}$. The elements of a group therefore fall into sets, in each of which the elements are conjugated to one another. Such sets are called *classes* of the group. A class is determined by specifying one of its elements; thus, knowing A, we can obtain the other elements by forming $G_bAG_b^{-1}$, where G_b runs through all the elements of the group (some elements of the class, however, may be repeated by this). By using the multiplication table for the group $\mathbf{C}_{3v}$ it is not difficult to satisfy oneself that its six elements fall into three classes: E; C_3, and C_3^2; and $\sigma_v^{(1)}$, $\sigma_v^{(2)}$, and $\sigma_v^{(3)}$.

We quote without proof the following further properties of classes:

1. The unit element forms a class by itself.
2. Except for the unit class, classes do not form subgroups since they do not contain the unit element.
3. Every element of an Abelian group forms a class by itself.
4. The number of elements in a class is a divisor of the order of the group.

A.2 Representation of Groups

A.2.1 Definition

A group of square matrices which is homomorphic with a given group is said to form a *representation* of the group. The number of rows or columns of the matrices is called the *dimension* of the representation.

It follows from this definition that one can associate with every element Q of a group a matrix $\Gamma(Q)$ such that corresponding to the product $QP = R$ of the group elements, one has the matrix product

$$\mathbf{\Gamma}(Q)\mathbf{\Gamma}(P) = \mathbf{\Gamma}(R). \tag{A.9}$$

Representations defined by Eq. (A.9) are called *vector* representations. A more general type of representation is possible, the matrices of which satisfy the relation

$$\mathbf{\Gamma}(Q)\mathbf{\Gamma}(P) = \varepsilon\mathbf{\Gamma}(R), \quad |\varepsilon| = 1. \tag{A.9a}$$

Representations such as these are known as *projective* or *ray* representations.

The multiplication of the matrices on the left-hand side of Eqs. (A.9) and (A.9a) is carried out according to the usual rules of matrix algebra. An element of the product matrix is given by

$$\Gamma_{ik}(QP) = \sum_{l} \mathbf{\Gamma}_{il}(Q)\mathbf{\Gamma}_{lk}(P). \tag{A.10}$$

If the matrices of a representation are all distinct, the representation is isomorphic with the group. Such a representation is then said to be *faithful.*

The matrices of a representation can be obtained by selection or by some other method so that they conform to the multiplication table for the group. In physical applications representations usually arise as the result of applying the elements of a group to functions of some coordinates. The groups which occur in physics are either groups of linear spatial transformations or of permutations of particle coordinates. When elements of these groups are applied to functions of the coordinates a set of new functions is generated, which transform linearly into one another under the action of the group elements. We now examine this process in somewhat greater detail.

Let ψ_0 be a coordinate function defined in the configuration space of the system.[2] The application of a element Q of a group $\mathbf{G}$ to ψ_0 converts this function into some other function which we denote by $\psi_Q \equiv Q\psi_0$. If Q runs through all the elements of the group, we obtain g functions. Some of these functions can be linearly

[2] For a system of N electrons the dimension of the configuration space is equal to $4N$. Each electron is described by three spatial coordinates and one spin coordinate.

dependent. Let the number of linearly independent functions ψ_i be equal to $f\ (f \le g)$. Under the action of the group elements the functions ψ_i transform only into each other, since by virtue of the properties of a group (see the theorem on page 136 of this appendix) not other functions may appear in this set. We thus have

$$Q\psi_k = \sum_{i=1}^{f} \Gamma_{ik}(Q)\psi_i. \tag{A.11}$$

The coefficients $\mathbf{\Gamma}_{ik}(Q)$ form a square matrix of order f. As follows from Eq. (A.11), under the action of the group element Q the function ψ_k transforms according to the kth column of the matrix $\mathbf{\Gamma}(Q)$. Each element Q of the group has a matrix $\mathbf{\Gamma}(Q)$ corresponding to it. Matrix corresponding to the product of two elements Q and P is a matrix, which is the product of the matrices $\mathbf{\Gamma}(Q)$ and $\mathbf{\Gamma}(P)$,

$$QP\psi_k = Q\sum_i \mathbf{\Gamma}_{ik}(P)\psi_i = \sum_l \left[\sum_i \mathbf{\Gamma}_{li}(Q)\mathbf{\Gamma}_{ik}(P)\right]\psi_l = \sum_l [\mathbf{\Gamma}(Q)\mathbf{\Gamma}(P)]_{lk}\psi_l.$$

The matrices $\mathbf{\Gamma}(Q)$ consequently form an f-dimensional representation of the group **G**. The set of f functions ψ_i, which define $\mathbf{\Gamma}(Q)$ matrices, are said to form a *basis* for the representation $\mathbf{\Gamma}$.

A.2.2 Vector Spaces

A vector space of n dimensions consists of the set of all the vectors $\mathbf{x}$ which can be obtained by forming all possible linear combinations of n linearly independent basis vectors $\mathbf{e}_i$:

$$\mathbf{x} = \sum_{i=1}^{n} x_i \mathbf{e}_i. \tag{A.12}$$

x_i is called the *component* of the vector $\mathbf{x}$ in the direction $\mathbf{e}_i$; x_i may also be a complex quantity, in which case the vector is complex.

Under a linear transformation from one system of basis vectors to another the components of a vector in the old system transform linearly into the components in the new system. Thus if

$$\mathbf{e}'_k = \sum_i s_{ik}\mathbf{e}_i, \tag{A.13}$$

then from

$$\mathbf{x} = \sum_i x_i \mathbf{e}_i = \sum_k x'_k \mathbf{e}'_k = \sum_{k,i} x'_k s_{ik} \mathbf{e}_i,$$

it follows that

$$x_i = \sum_k s_{ik} x'_k. \tag{A.14}$$

If we represent the vector $\mathbf{x}$ as a *column matrix* X and denote the square transformation matrix in (A.13) by S, then Eq. (A.14) can be written as

$$X = SX', \tag{A.15}$$

From which

$$X' = S^{-1}X, \tag{A.16}$$

where S^{-1} is the matrix of the inverse transformation.

Instead of transforming the basis vectors, we can carry out a transformation of the vector space by which we associate with each vector $\mathbf{x}$ a vector $\mathbf{y}$:

$$\mathbf{y} = A\mathbf{x}. \tag{A.17}$$

The quantity A can be regarded as an operator which carries the vector $\mathbf{x}$ into vector $\mathbf{y}$. The analytical form of A is an n-dimensional matrix, which connects the components of the vectors:

$$y_i = \sum_k a_{ik} x_k, \tag{A.18}$$

or in a matrix notation

$$Y = AX. \tag{A.19}$$

Let the vectors $\mathbf{y}$ and $\mathbf{x}$ be connected to each other by the relation (A.19). We carry out the transformation (A.13) of the basis vectors. Thus, in the new coordinate system

$$Y' = A'X'. \tag{A.20}$$

Let us now find the relation between A' and A. For this purpose we pass to the old coordinate system, carry out in it the transformation (A.19), and return to the new system:

$$Y' = S^{-1}Y, \quad Y = AX, \quad X = SX'$$

or

$$Y' = S^{-1}ASX'. \tag{A.21}$$

On comparing (A.20) and (A.21), we obtain

$$A' = S^{-1}AS. \tag{A.22}$$

If the basis vectors form an orthogonal set

$$\mathbf{e}_i \cdot \mathbf{e}_j = \delta_{ij}, \tag{A.23}$$

then the scalar product of two vectors is given by

$$\mathbf{x} \cdot \mathbf{y} = \sum_i x_i^* y_i, \tag{A.24}$$

and the square of the length of a vector is equal to the sum of the squared moduli of its components:

$$\mathbf{x} \cdot \mathbf{x} = |\mathbf{x}|^2 = \sum_i |x_i|^2. \tag{A.25}$$

A linear transformation is said to be *unitary*, if it leaves unchanged the scalar product of two vectors. It follows from this definition of a unitary transformation that

$$\sum_k x_k'^* y_k' = \sum_i x_i^* y_i = \sum_{i,k,l} s_{ik}^* s_{il} x_k'^* y_l'. \tag{A.26}$$

Eq. (A.26) holds if the condition

$$\sum_i s_{ik}^* s_{il} = \delta_{kl}. \tag{A.27}$$

is fulfilled. However, according to the definition of inverse of a matrix,

$$\sum_i s_{ki}^{-1} s_{il} = \delta_{kl}, \tag{A.28}$$

and it therefore follows from a comparison of (A.27) and (A.28) that $s_{ki}^{-1} = s_{ik}^*$. This can be symbolically written in the form[3]

$$S^{-1} = \tilde{S}^* \tag{A.29}$$

or

$$S\tilde{S}^* = \tilde{S}^* S = I. \tag{A.29a}$$

Matrices, which satisfy the condition (A.29) or (A.29a), are called *unitary*. From (A.29a) it can be seen that in addition to (A.27) the elements of a unitary matrix also obey the relation

$$\sum_k s_{ik}^* s_{lk} = \delta_{il}. \tag{A.27a}$$

In the case of real matrices we have instead of (A.29a)

$$S\tilde{S} = \tilde{S}S = I. \tag{A.30}$$

Matrices which satisfy (A.30) are said to be *orthogonal*. Their elements satisfy the following orthogonality conditions:

$$\sum_i s_{ik} s_{il} = \delta_{kl}, \quad \sum_k s_{ik} s_{lk} = \delta_{il}. \tag{A.31}$$

A.2.3 Reducibility of Representations

The f basis functions for a representation $\mathbf{\Gamma}$ can be regarded as basis vectors in an f-dimensional vector space. We denote this space by $\mathfrak{R}$ and say that it transforms according to the representation $\mathbf{\Gamma}$. According to Eq. (A.11), the effect of an element Q of a group $\mathbf{G}$ upon a basis vector ψ_k is to generate a vector $Q\psi_k$ which also lies in $\mathfrak{R}$. Since any vector in a vector space can be expressed in the form of a linear combination of the basis vectors, the effect on an element of the group is to convert every vector of the space into another vector which also belongs to $\mathfrak{R}$. The space $\mathfrak{R}$ is therefore invariant under the transformations of the group $\mathbf{G}$.

[3] $\tilde{S}$ denotes the transpose of the matrix S. Its elements are related to those of S by $\tilde{S}_{ik} = S_{ki}$

According to Eq. (A.22) a linear transformation (A.13) of the basis vectors,

$$\psi_k' = \sum_{i=1}^{f} s_{ik}\psi_i, \tag{A.32}$$

converts the matrices of a representation $\mathbf{\Gamma}$ into the matrices

$$\mathbf{\Gamma}'(Q) = S^{-1}\mathbf{\Gamma}(Q)S. \tag{A.33}$$

Two representations which are connected by a relation such as (A.33) are said to be *equivalent*. It is clear that there is an infinite number of equivalent representations. A transformation, which converts a representation into an equivalent one, is called a *similarity transformation*.

If a similarity transformation can be found, which converts all the matrices of a representation into block-diagonal form

$$\mathbf{\Gamma}' = S^{-1}\,\mathbf{\Gamma}S = \begin{vmatrix} \mathbf{\Gamma}^{(1)} & & & \\ & \mathbf{\Gamma}^{(2)} & & \\ & & \cdots & \\ & & & \mathbf{\Gamma}^{(m)} \end{vmatrix} \tag{A.34}$$

then the representation is said to be *reducible*. As a result of such a similarity transformation, the representation is decomposed into m representations of smaller dimensions. This is written as follows:

$$\mathbf{\Gamma} \doteq \mathbf{\Gamma}^{(1)} + \mathbf{\Gamma}^{(2)} + \cdots + \mathbf{\Gamma}^{(m)}. \tag{A.35}$$

The reducibility of a representation implies that by means of a linear transformation of the basis vector the space $\mathfrak{R}$ can be divided up into a number of invariant subspaces $\mathfrak{R}^{(\alpha)}$ each of which transform according to a representation $\mathbf{\Gamma}^{(\alpha)}$. The operations of **G** transform the vector of each subspace among themselves without mixing vectors from different subspaces.

If there is no transformation which reduces the representation matrices to the block-diagonal form (A.34), then the representation is said to be *irreducible*. We note that an irreducible representation of a group obviously forms a representation of a subgroup of the group. However, as far as the subgroup is concerned, this representation may be reducible and may be decomposed into irreducible representations of the subgroup. This process is called *reduction with respect to a subgroup*.

The following set of matrices offers an example of reducible representation of the groups $\mathbf{C}_{3v}$:

$$\boldsymbol{\Gamma}(E)=\begin{vmatrix}1&0&0\\0&1&0\\0&0&1\end{vmatrix},\quad \boldsymbol{\Gamma}(C_3)=\begin{vmatrix}0&0&1\\1&0&0\\0&1&0\end{vmatrix},\quad \boldsymbol{\Gamma}(C_3^2)=\begin{vmatrix}0&1&0\\0&0&1\\1&0&0\end{vmatrix},$$

$$\boldsymbol{\Gamma}\left(\sigma_v^{(1)}\right)=\begin{vmatrix}1&0&0\\0&0&1\\0&1&0\end{vmatrix},\ \boldsymbol{\Gamma}\left(\sigma_v^{(2)}\right)=\begin{vmatrix}0&0&1\\0&1&0\\1&0&0\end{vmatrix},\ \boldsymbol{\Gamma}\left(\sigma_v^{(3)}\right)=\begin{vmatrix}0&1&0\\1&0&0\\0&0&1\end{vmatrix}. \tag{A.36}$$

It is easily verified by direct multiplication that these matrices conform to the multiplication table for the group (see Table A.1). By means of a transformation such as (A.34) it can be decomposed into two irreducible representations, one of which is one-dimensional and the other is two-dimensional. In the following sections we shall describe method of the decomposition of reducible representations and, in particular, will obtain the decomposition (A.35) of the representation (A.36).

A.2.4 *Properties of Irreducible Representations*

We quote without proof the following important properties of irreducible representation of finite groups.[4]

1. The number of nonequivalent irreducible representations of a group is equal to the number of classes in the group.
2. The sum of the squares of the dimensions of the nonequivalent irreducible representation is equal to the order of the group, that is,

$$f_1^2+f_2^2+\cdots+f_r^2=g, \tag{A.37}$$

   where f_α denotes the dimension of the α-th irreducible representation. It follows from this that all the irreducible representations of an Abelian group are one-dimensional, since the number of its irreducible representations is equal to the number of elements in the group.

[4] The proof of these statements can be found in any text on group theory, for example, in Refs. [3, 4].

3. The dimension of an irreducible representation of a finite group is a divisor of the order of the group.
4. The following orthogonality relations hold for the matrix elements of irreducible representations:

$$\sum_{R} \Gamma_{ik}^{(\alpha)}(R)^* \Gamma_{lm}^{(\beta)}(R) = \left(\frac{g}{f_\alpha}\right) \delta_{\alpha\beta}\delta_{il}\delta_{km}, \tag{A.38}$$

$$\sum_{\alpha,i,k} \left(\frac{f_\alpha}{g}\right) \Gamma_{ik}^{(\alpha)}(R)^* \Gamma_{ik}^{(\alpha)}(Q) = \delta_{RQ}. \tag{A.39}$$

The summation in (A.38) is taken over all the g elements of the group. In (A.39) all f_α^2 elements of the matrix $\mathbf{\Gamma}^{(\alpha)}(R)$ are summed for each irreducible representation $\mathbf{\Gamma}^{(\alpha)}$. The number of terms in the sum (A.39) is equal to the order of the group in accordance with Eq. (A.37).

A.2.5 Characters

Let the matrices $\mathbf{\Gamma}^{(\alpha)}(R)$ form a representation of a group $\mathbf{G}$. The sum of the diagonal elements of a matrix $\mathbf{\Gamma}^{(\alpha)}(R)$ is called the *character* of the element R in the representation $\mathbf{\Gamma}^{(\alpha)}$ and is denoted by $\chi^{(\alpha)}(R)$:

$$\chi^{(\alpha)}(R) = \sum_{i} \Gamma_{ii}^{(\alpha)}(R). \tag{A.40}$$

Each representation is thus characterized by a set of g characters.

The characters of equivalent representations are identical, for in accordance with the rule for matrix multiplication,

$$\sum_{i} \left(S^{-1}\mathbf{\Gamma}(R)S\right)_{ii} = \sum_{i,l,m} s_{il}^{-1} \Gamma_{lm}(R) s_{mi}$$

$$= \sum_{l,m} \left(\sum_{i} s_{mi} s_{il}^{-1}\right) \Gamma_{lm}(R) = \sum_{l,m} \delta_{ml} \Gamma_{lm}(R) = \sum_{m} \Gamma_{mm}(R).$$

Hence the specification of a representation by means of a set of characters does not distinguish between equivalent representations. This is extremely useful, since in physical applications only inequivalent representations are important. By specifying the characters of a representation one can, of course, still distinguish nonequivalent representations.

The elements of a group belonging to a given class are connected among themselves by relations analogous to (A.33). Their characters must therefore all be identical. Consequently, one can denote a class by specifying the character of just one of its members. Furthermore, the number of distinct characters of a representation cannot exceed the number of classes in the group.

The characters of irreducible representations satisfy the following orthogonality relations:

$$\sum_{R} \chi^{(\alpha)}(R)^{*}\chi^{(\beta)}(R) = g\delta_{\alpha\beta}. \tag{A.41}$$

These relations are obtained from Eq. (A.38) by putting i equal to k and l equal to m and summing over k and m on both sides.

Since the characters of all the elements belonging to a given class are equal, Eq. (A.41) can be written in the form

$$\sum_{\mathbf{C}} g_{\mathbf{C}}\chi^{(\alpha)}(\mathbf{C})^{*}\chi^{(\beta)}(\mathbf{C}) = g\delta_{\alpha\beta}, \tag{A.42}$$

where the sum is taken over all classes $\mathbf{C}$ of the group, and $g_{\mathbf{C}}$ denotes the number of elements in $\mathbf{C}$. It is evident that $\sum_{C} g_{C} = g$.

If we denote $(g_C/g)^{1/2}\chi^{(\alpha)}(C)$ as a_{C_α}, then since number of classes is equal to the number of irreducible representations, the quantities a_{C_α} form a square matrix. The orthogonality relations for a_{C_α}, Eq. (A.42), are then identical with the unitary condition (A.27) for the first subscript. The matrix $\|a_{\mathrm{C}_\alpha}\|$ is therefore unitary, with the unitary condition (A.27a) also applying to its elements. From this we obtain a second orthogonality relation for the characters of irreducible representations:

$$\sum_{\alpha} \chi^{(\alpha)}(\mathbf{C}_i)^{*}\chi^{(\alpha)}(\mathbf{C}_k) = \left(\frac{g}{g_{\mathbf{C}_i}}\right)\delta_{\mathbf{C}_i\mathbf{C}_k}. \tag{A.43}$$

Among the irreducible representations of a group there is always a one-dimensional representation which is generated by a basis function that is invariant under all the operations of the group. All the characters of this representation are equal to unity. This representation is usually called the *totally symmetric* or *unit* representation, and is denoted by the symbol A_1.

A.2.6 The Decomposition of a Reducible Representation

Consider some reducible representation $\mathbf{\Gamma}$. By definition, one can always find a unitary transformation that will bring it to the block-diagonal form (A.34). We assume that no further reduction is possible, that is, the representations $\mathbf{\Gamma}^{(\alpha)}$, which appear in the decomposition (A.35), are irreducible. A given irreducible representations, however, may occur several times in the decomposition. Thus, in general we have

$$\boldsymbol{\Gamma} \doteq \sum_{\beta} a^{(\beta)} \boldsymbol{\Gamma}^{(\beta)}, \tag{A.44}$$

where $a^{(\beta)}$ denotes the number of times the representation $\boldsymbol{\Gamma}^{(\beta)}$ occurs in the decomposition of $\boldsymbol{\Gamma}$. Now, since a similarity transformation does not alter the characters of a representation, the characters of the representation $\boldsymbol{\Gamma}$ are given by

$$\chi^{(\Gamma)}(R) = \sum_{\beta} a^{(\beta)} \chi^{(\beta)}(R). \tag{A.45}$$

We multiply (A.45) throughout by $\chi^{(\alpha)}(R)^*$ and sum over all the elements in the group. By virtue of the orthogonality relations (A.41) we obtain

$$a^{(\alpha)} = \left(\frac{1}{g}\right) \sum_{R} \chi^{(\Gamma)}(R) \chi^{(\alpha)}(R)^*. \tag{A.46}$$

For actual applications it is convenient to rewrite this relation in the form

$$a^{(\alpha)} = \left(\frac{1}{g}\right) \sum_{\mathbf{C}} g_{\mathbf{C}} \chi^{(\Gamma)}(\mathbf{C}) \chi^{(\alpha)}(\mathbf{C})^*. \tag{A.47}$$

If the characters of the representations are known, Eq. (A.47) enables one to determine easily the irreducible representations, which occur in the decomposition of a given reducible representation.

As an example we determine the irreducible representations of $\mathbf{C}_{3v}$, which occur in the decomposition of the representation (A.36). The characters of this representation are

	E	$2C_3$	$3\sigma_v$
$\chi^{(\Gamma)}$	3	0	1

The characters of the irreducible representations of the point group $\mathbf{C}_{3v}$ are represented in Table A.2. We find from formula (A.47) that

$$a^{(1)} = \frac{1}{6}(3+3) = 1, \quad a^{(2)} = \frac{1}{6}(3-3) = 0, \quad a^{(3)} = \frac{1}{6}6 = 1.$$

Table A.2 Characters of the irreducible representations of C_{3v}

	E	$2C_3$	$3\sigma_v$
$\chi^{(1)}$	1	1	1
$\chi^{(2)}$	1	1	−1
$\chi^{(3)}$	2	−1	0

The representation (A.36) consequently breaks down into two irreducible representations: the one-dimensional representation $\mathbf{\Gamma}^{(1)}$ and the two-dimensional representation $\mathbf{\Gamma}^{(3)}$.

A.2.7 The Direct Product of Representations

Let us consider two irreducible representations $\mathbf{\Gamma}^{(\alpha)}$ and $\mathbf{\Gamma}^{(\beta)}$ of a group **G**, each of which is defined by a set of basis functions $\psi_i^{(\alpha)}(i = 1, 2, \ldots, f_\alpha)$ and $\psi_k^{(\beta)}(k = 1, 2, \ldots, f_\beta)$, respectively. If we construct all possible products of the form $\psi_i^{(\alpha)}\,\psi_k^{(\beta)}$, we obtain an $f_\alpha f_\beta$-dimensional basis for a representation of the group. This representation is known as the *direct product* of the representations $\mathbf{\Gamma}^{(\alpha)}$ and $\mathbf{\Gamma}^{(\beta)}$ and is denoted by $\mathbf{\Gamma}^{(\alpha)} \times \mathbf{\Gamma}^{(\beta)}$. The matrix elements of the direct product of the two representations are of the form of products of the matrix elements of $\mathbf{\Gamma}^{(\alpha)}$ and $\mathbf{\Gamma}^{(\beta)}$,

$$R\psi_i^{(\alpha)}\psi_k^{(\beta)} = \left(R\psi_i^{(\alpha)}\right)\left(R\psi_k^{(\beta)}\right) = \sum_{l,m} \Gamma_{li}^{(\alpha)}(R)\Gamma_{mk}^{(\beta)}(R)\psi_l^{(\alpha)}\psi_m^{(\beta)}, \tag{A.48}$$

and the characters of this representation are equal to the product of the constituent characters:

$$\chi^{(\alpha\times\beta)}(R) = \sum_{i,k} \Gamma_{ii}^{(\alpha)}(R)\Gamma_{kk}^{(\beta)}(R) = \chi^{(\alpha)}(R)\chi^{(\beta)}(R), \tag{A.49}$$

The matrix $\mathbf{\Gamma}^{(\alpha\times\beta)}(R)$, with matrix elements given by (A.48), is called the direct product of the matrices $\mathbf{\Gamma}^{(\alpha)}(R)$ and $\mathbf{\Gamma}^{(\beta)}(R)$.[5] In contrast to the ordinary product of square matrices, where the order of the resulting matrices is the same as that of the matrices that are multiplied, the order of the matrix resulting from the direct product of two matrices is equal to the product of the orders of the matrices that are multiplied. The direct product of two matrices can be represented in a block form; thus for matrices of order 2

[5] The equivalent term "Kronecker product of matrices" is also used in the literature.

$$A \times B = \begin{vmatrix} a_{11}B & a_{12}B \\ a_{21}B & a_{22}B \end{vmatrix} = \begin{vmatrix} a_{11}b_{11} & a_{11}b_{12} & a_{12}b_{11} & a_{12}b_{12} \\ a_{11}b_{21} & a_{11}b_{22} & a_{12}b_{21} & a_{12}b_{22} \\ a_{21}b_{11} & a_{21}b_{12} & a_{22}b_{11} & a_{22}b_{12} \\ a_{21}b_{21} & a_{21}b_{22} & a_{22}b_{21} & a_{22}b_{22} \end{vmatrix}. \tag{A.50}$$

It is clear that although $A \times B \neq B \times A$, the two can be converted into each other by an appropriate permutation of the rows and columns. The characters of a direct product are therefore independent of the order in which the representations are multiplied. This also follows from formula (A.49).

If the two representations that are multiplied together are identical but possess different bases, we have:

$$\chi^{(\alpha \times \alpha)}(R) = \left[\chi^{(\alpha)}(R)\right]^2. \tag{A.51}$$

Thus, the characters of the direct product of the representations $\Gamma^{(3)} \times \Gamma^{(3)}$ of $\mathbf{C}_{3v}$ are (see Table A.2)

	E	$2C_3$	$3\sigma_v$
$[\chi^{(3)}(R)]^2$	4	1	0

We now reduce this representation. Its decomposition into irreducible components is carried out according to the formula (A.47) and contains each irreducible representation of $\mathbf{C}_{3v}$ once, that is,

$$\boldsymbol{\Gamma}^{(3)} \times \boldsymbol{\Gamma}^{(3)} \doteq \boldsymbol{\Gamma}^{(1)} + \boldsymbol{\Gamma}^{(2)} + \boldsymbol{\Gamma}^{(3)}. \tag{A.52}$$

If in the direct product $\Gamma^{(\alpha)} \times \Gamma^{(\alpha)}$ the bases of the two representations $\Gamma^{(\alpha)}$ coincide, then the basis of the direct product is symmetric with respect to a permutation of its two factors. A direct product of this kind is called the *symmetric product* of a basis with itself, and is denoted by $[\Gamma^{(\alpha)}]^2$. Its dimension is less than $f_\alpha^{\,2}$ and is equal to $f_\alpha(f_\alpha + 1)/2$; the characters of a symmetric product are not given by a product of characters (A.51), but by an equation, which is a special case of the more general formula, see Ref. [1], Eq. (4.18):

$$\left[\chi^{(\alpha)}\right]^2(R) = \frac{1}{2}\chi^{(\alpha)}(R^2) + \frac{1}{2}\left[\chi^{(\alpha)}(R)\right]^2. \tag{A.53}$$

For example, the characters of the symmetric product $[\Gamma^{(3)}]^2$ of $\mathbf{C}_{3v}$ are

	E	$2C_3$	$3\sigma_v$
$[\chi^{(3)}]^2(R)$	3	0	1

Decomposing this representation into its irreducible components, we obtain

$$\left[\mathbf{\Gamma}^{(3)}\right]^2 \doteq \mathbf{\Gamma}^{(1)} + \mathbf{\Gamma}^{(3)}. \tag{A.54}$$

The direct product of two irreducible representations is always reducible except when one of the representations is one-dimensional. The decomposition of a direct product into irreducible representations

$$\mathbf{\Gamma}^{(\alpha)} \times \mathbf{\Gamma}^{(\beta)} \doteq \sum_{\tau} a^{(\tau)} \mathbf{\Gamma}^{(\tau)}, \tag{A.55}$$

is conventionally called the *Clebsch–Gordan series*. The coefficients in the decomposition are found from the general formula (A.46):

$$a^{(\tau)} = \left(\frac{1}{g}\right) \sum_{R} \chi^{(\alpha\times\beta)}(R)\chi^{(\tau)}(R)^* = \left(\frac{1}{g}\right) \sum_{R} \chi^{(\alpha)}(R)\chi^{(\beta)}(R)\chi^{(\tau)}(R)^*. \tag{A.56}$$

We now determine the condition that the totally symmetric representation A_1 appears in the decomposition (A.55). The characters of this representation $\chi^{(A_1)}(R) = 1$ for all elements in the group, and hence by virtue of the orthogonality relations (A.41) we obtain

$$a^{(A_1)} = \left(\frac{1}{g}\right) \sum_{R} \chi^{(\alpha)}(R)\chi^{(\beta)}(R) = \delta_{\beta\alpha^*}, \tag{A.57}$$

where α^* in the right-hand side of Eq. (A.57) denotes the representation whose matrix elements are the complex conjugates of the matrix elements of $\mathbf{\Gamma}^{(\alpha)}$. Such representations are called *complex conjugate representations*. Consequently, the totally symmetric representation occurs in the decomposition of a direct product of irreducible representations if, and only if, these representations are the complex conjugates of one another. In the case of real matrices the totally symmetric representation occurs only in the direct product of an irreducible representation with itself.

In analogy with the direct product of two irreducible representations of a group, one may define a direct product of an arbitrary number of irreducible representations.

The characters of such a direct product are equal to the product of the characters of the representations that are multiplied:

$$\chi^{(\alpha\times\beta\times\cdots\times\omega)}(R)=\chi^{(\alpha)}(R)\chi^{(\beta)}(R)\ldots\chi^{(\omega)}(R). \tag{A.58}$$

The direct products of representations that have been considered so far concern one and the same group. If $\mathbf{G}_1\times\mathbf{G}_2$ is the direct product of two groups (see section 1.7 in Ref. [1]), and the irreducible representations $\mathbf{\Gamma}^{(\alpha)}\in\mathbf{G}_1$, $\mathbf{\Gamma}^{(\alpha)}\in\mathbf{G}_2$, then the direct product of the representations $\mathbf{\Gamma}^{(\alpha)}\times\mathbf{\Gamma}^{(\beta)}$ is an irreducible representation of the group $\mathbf{G}_1\times\mathbf{G}_2$. By a derivation similar to that which led to (A.49), it can be shown that the element $R=R_1R_2$ of the group $\mathbf{G}_1\times\mathbf{G}_2$ possesses the character

$$\chi^{(\alpha\times\beta)}(R)=\chi^{(\alpha)}(R_1)\chi^{(\beta)}(R_2). \tag{A.59}$$

The different irreducible representations of $\mathbf{G}_1\times\mathbf{G}_2$ are obtained by combining in pairs the irreducible representations of $\mathbf{G}_1$ and $\mathbf{G}_2$.

A.2.8 Clebsch–Gordan Coefficients

The reduction of a direct product of representations into its irreducible components (A.55) is accomplished by means of a linear transformation of the basis functions $\psi_i^{(\alpha)}\psi_i^{(\beta)}$ to a set of functions $\psi_t^{(\tau)}$ which do not mix under the operation of the group elements:

$$\psi_t^{(a\tau)}=\sum_{i,k}\psi_i^{(\alpha)}\psi_k^{(\beta)}\langle\alpha i,\,\beta k|a\tau t\rangle. \tag{A.60}$$

The index a distinguishes the representations τ should they occur more than once in the decomposition (A.55). The coefficients $\langle\alpha i,\beta k|a\tau t\rangle$ appearing in this equation are known as *Clebsch–Gordan coefficients*. Since the basis functions are normally chosen to be orthogonal, the Clebsch–Gordan coefficients constitute a unitary matrix[6] which we denote by $C_{\alpha\beta}$. This matrix reduces the direct product. Thus, in agreement with (A.34)

[6] Basis functions $\psi_t^{(a\tau)}$ that belong to different irreducible representations are automatically orthogonal to one another (see Section A.2.10). Even when an irreducible representation is repeated in the decomposition (A.55), its bases can always be orthogonalized to one another.

$$C_{\alpha\beta}^{-1}(\mathbf{\Gamma}^{(\alpha)}\times\mathbf{\Gamma}^{(\beta)})C_{\alpha\beta}=\begin{pmatrix}\mathbf{\Gamma}^{(\tau_1)} & & & \\ & \mathbf{\Gamma}^{(\tau_2)} & & \\ & & \cdots & \\ & & & \mathbf{\Gamma}^{(\tau_3)}\end{pmatrix} \tag{A.61}$$

The rows of the $C_{\alpha\beta}$ matrix are numbered by the indices i and k, which take on a total of $f_\alpha f_\beta$ values, and the columns are numbered by the indices a, τ, and t. For most of the groups that occur in physics one may choose the matrix of Clebsch–Gordan coefficients to be real, so that the orthogonality relations (A.31) apply to its elements:

$$\begin{aligned}&\sum_{a,\tau,t}\langle\alpha i,\beta k|a\tau t\rangle\langle\alpha\bar{i},\beta\bar{k}|a\tau t\rangle=\delta_{i\bar{i}}\delta_{k\bar{k}},\\&\sum_{i,k}\langle\alpha i,\beta k|a\tau t\rangle\langle\alpha i,\beta k|\bar{a}\bar{\tau}\bar{t}\rangle=\delta_{a\bar{a}}\delta_{\tau\bar{\tau}}\delta_{t\bar{t}}.\end{aligned} \tag{A.62}$$

In an orthogonal transformation the inverse of a matrix is equal to its transpose, and therefore the inverse transformation to (A.60) is

$$\psi_i^{(\alpha)}\psi_k^{(\beta)}=\sum_{a,\tau,t}\psi_t^{(a\tau)}\langle a\tau t|\alpha i,\beta k\rangle, \tag{A.63}$$

or in matrix form

$$\mathbf{\Gamma}^{(\alpha)}\times\mathbf{\Gamma}^{(\beta)}=C_{\alpha\beta}\left(\sum_{a,\tau}\mathbf{\Gamma}^{(a\tau)}\right)C_{\alpha\beta}^{-1}. \tag{A.64}$$

Eq. (A.64) is equivalent to an equation between matrix elements:

$$\Gamma_{i\bar{i}}^{(\alpha)}(R)\Gamma_{k\bar{k}}^{(\beta)}(R)=\sum_{a,\tau}\sum_{t,\bar{t}}\langle\alpha i,\beta k|a\tau t\rangle\Gamma_{t\bar{t}}^{(a\tau)}(R)\langle a\tau\bar{t}|\alpha\bar{i},\beta\bar{k}\rangle. \tag{A.65}$$

If irreducible representations occur more than once in the decomposition (A.55), the matrix of Clebsch–Gordan coefficients is defined only within a unitary transformation. This case has been considered by Koster [5].

From the definition of Clebsch–Gordan coefficients it follows that

$$\langle\alpha i,\beta k|\tau t\rangle\equiv0 \tag{A.66}$$

for all $\mathbf{\Gamma}^{(\tau)}$ that do not appear in the decomposition of the direct product $\mathbf{\Gamma}^{(\alpha)}\times\mathbf{\Gamma}^{(\beta)}$.

A.2.9 *The Regular Representation*

Let ψ_0 be some function of coordinates, which does not possess any symmetry properties with respect to the operations of a group **G**. The operation of the g elements of the group upon ψ_0 generates g linearly independent functions

$$\psi_R = R\psi_0. \tag{A.67}$$

The functions ψ_R are converted into one another under the operations of the group,

$$Q\psi_R = QR\psi_0 = P\psi_0 \equiv \psi_P. \tag{A.68}$$

The functions (A.67) therefore constitute a g-dimensional basis for a representation of **G**. A representation such as this is called a *regular representation*. According to (A.68), the matrices of the regular representation, with the exception of the matrix of the identity transformation E, possess zero diagonal elements. The identity matrix is always a diagonal unit matrix. From this it follows that the characters of the regular representation are given by

$$\chi(E) = g, \quad \chi(R) = 0 \quad \text{for} \quad R \neq E. \tag{A.69}$$

Substituting (A.69) into (A.46) and remembering that for any representation $\chi^{(\alpha)}(E) = f_\alpha$, we find that

$$a^{(\alpha)} = \left(\frac{1}{g}\right)(f_\alpha g) = f_\alpha,$$

that is, the number of times each irreducible representation occurs in the decomposition of the regular representation is equal to its dimension.

Let us write formula (A.45), which is an expression for the character of a reducible representation in terms of the irreducible representations occurring in it, for the case $R = E$:

$$\sum_\beta a^{(\beta)} f_\beta = f_\Gamma.$$

If the reducible representation is the regular representation, $f_\Gamma = g$ and $a^{(\beta)} = f_\beta$. Hence, we arrive at the result that

$$\sum_\beta f_\beta^2 = g. \tag{A.70}$$

We have thus proved a property of which we have already made use: the sum of the squares of the dimensions of all the irreducible representations of a group is equal to the order of the group.

We note that in order to construct the regular representation, one does not have to make use of the basis functions (A.67). The elements of the group themselves can be employed as a basis. A knowledge of the multiplication table of the group is therefore sufficient to write down all the matrices of the regular representation.

A.2.10 The Construction of Basis Functions for Irreducible Representation

The regular representation considered in the previous section is generated by the g linearly independent functions (A.67). The decomposition of this representation into its irreducible components contains all the nonequivalent irreducible representations of the group. We show that this decomposition can be accomplished by constructing the following linear combinations of the basis functions ψ_R:

$$\psi_{ik}^{(\alpha)} = \left(\frac{f_\alpha}{g}\right)\sum_R \Gamma_{ik}^{(\alpha)}(R)^* \psi_R, \tag{A.71}$$

where $\Gamma_{ik}^{(\alpha)}(R)$ is a matrix element of the irreducible representation $\Gamma^{(\alpha)}$ corresponding to the operation R. The summation is taken over all g operations in the group.

We apply an arbitrary operation Q of the group to the function (A.71)

$$Q\psi_{ik}^{(\alpha)} = \left(\frac{f_\alpha}{g}\right)\sum_R \Gamma_{ik}^{(\alpha)}(R)^* QR\psi_0 = \left(\frac{f_\alpha}{g}\right)\sum_P \Gamma_{ik}^{(\alpha)}(Q^{-1}P)^* P\psi_0. \tag{A.72}$$

In this equation we have denoted the operation QR by P and made use of the invariance property of a sum over the group, Eq. (A.7). Then we write the matrix element of the product as products of matrix elements and make use of the property (A.29) of unitary matrices:

$$\Gamma_{ik}^{(\alpha)}(Q^{-1}P)^* = \sum_l \Gamma_{il}^{(\alpha)}(Q^{-1})^* \Gamma_{lk}^{(\alpha)}(P)^* = \sum_l \Gamma_{li}^{(\alpha)}(Q)\Gamma_{lk}^{(\alpha)}(P)^*. \tag{A.73}$$

Substituting (A.73) into (A.72), we obtain finally

$$Q\psi_{ik}^{(\alpha)} = \sum_l \Gamma_{li}^{(\alpha)}(Q)\psi_{lk}^{(\alpha)}. \tag{A.74}$$

Hence the function $\psi_{ik}^{(\alpha)}$ transforms as the ith column of the irreducible representation $\Gamma^{(\alpha)}$, and the set of f_α functions $\psi_{ik}^{(\alpha)}$ with fixed second index k forms a basis for the irreducible representation $\Gamma^{(\alpha)}$. One can form altogether f_α independent bases corresponding to the number of different values for k. This is to be expected, since in the decomposition of regular representation, each irreducible representation occurs as many times as its dimension.

From the form of Eq. (A.71) it follows that in order to obtain a basis function for a representation $\Gamma^{(\alpha)}$, it is sufficient to apply the operator

$$\varepsilon_{ik}^{(\alpha)} = \left(\frac{f_\alpha}{g}\right)\sum_R \Gamma_{ik}^{(\alpha)}(R)^* R, \tag{A.75}$$

to some arbitrary functions ψ_0. There are f_α^2 such operators for every irreducible representation. They form f_α sets which differ from one another in the second index. Each of these sets can be used to obtain basis functions for an irreducible representation.

If the function ψ_0 possesses specific symmetry properties with respect to the operations of the group **G**, then in some cases the application of the operator $\varepsilon_{ik}^{(\alpha)}$ to it may give zero. Thus let us determine the result of applying $\varepsilon_{ik}^{(\alpha)}$ to a function of the form (A.71)

$$\begin{aligned} \varepsilon_{ik}^{(\alpha)}\psi_{mn}^{(\beta)} &= \left(\frac{f_\alpha}{g}\right)\sum_R \Gamma_{ik}^{(\alpha)}(R)^* R\psi_{mn}^{(\beta)} \\ &= \left(\frac{f_\alpha}{g}\right)\sum_l \sum_R \Gamma_{ik}^{(\alpha)}(R)^*\Gamma_{lm}^{(\beta)}(R)\psi_{ln}^{(\beta)} = \delta_{\alpha\beta}\delta_{km}\psi_{in}^{(\alpha)}. \end{aligned} \tag{A.76}$$

In this equation we made use of (A.74) and the orthogonality relations (A.38). The application of $\varepsilon_{ik}^{(\alpha)}$ to a basis function of an irreducible representation therefore either gives zero or another basis function belonging to the same irreducible representation. When $i = k$, the application of $\varepsilon_{ii}^{(\alpha)}$ to a basis function $\psi_{ik}^{(\alpha)}$ gives just the same function once more. Operators which possess such properties are known as *projection operators*. They satisfy the operator equation

$$\varepsilon_{ii}^{(\alpha)}\varepsilon_{ii}^{(\alpha)} = \varepsilon_{ii}^{(\alpha)}. \tag{A.77}$$

Let the function $\psi_{ik}^{(\alpha)}$ transform according to the ith column of a representation $\Gamma^{(\alpha)}$, the index k specifying the way in which the functions in this basis are constructed (in a particular case, the $\psi_{ik}^{(\alpha)}$ can be constructed by Eq. (A.71), although this is by no means necessary for the argument that follows). We now show that functions which transform according to different irreducible representations, or according to different columns of the same representation, are orthogonal to one another, that is,

$$\Omega = \int \psi_{ik}^{(\alpha)^*} \psi_{mn}^{(\beta)} dV = \delta_{\alpha\beta}\delta_{im} \begin{cases} A(\alpha;k,n) & \text{for } k \neq n, \\ 1 & \text{for } k = n, \end{cases} \tag{A.78}$$

where $A(\alpha; k, n)$ is determined by the choice of the bases k and n, but is independent of i and m.

To prove this, we make use of the fact that an integral over all space is invariant under any transformation of the coordinates. The integral (A.78) is therefore unchanged if the integrand is operated upon by some operation of the group:

$$\Omega = \int R\psi_{ik}^{(\alpha)^*} R\psi_{mn}^{(\beta)} dV = \sum_{\mu,\nu} \Gamma_{\mu i}^{(\alpha)}(R)^* \Gamma_{\nu m}^{(\beta)}(R) \int \psi_{\mu k}^{(\alpha)^*} \psi_{\nu n}^{(\beta)} dV. \tag{A.79}$$

We sum this equation over all the operations in the group. The integral on the left-hand side is then simply multiplied by the order of the group. Applying the orthogonality relations (A.38) to the expression on the right-hand side, we obtain as a result

$$g\Omega = \delta_{\alpha\beta}\delta_{im} \left(\frac{g}{f_\alpha}\right) \sum_{\mu} \int \psi_{\mu k}^{(\alpha)^*} \psi_{\mu n}^{(\alpha)} dV. \tag{A.80}$$

The sum over μ in (A.80) is independent of i and m. If $k = n$, it is equal to f_α, due to the orthonormality of the basis functions. When $k \neq n$ we put

$$\left(\frac{1}{f_\alpha}\right) \sum_{\mu} \int \psi_{\mu k}^{(\alpha)^*} \psi_{\mu n}^{(\alpha)} dV = A(\alpha;k,n),$$

and so arrive at Eq. (A.78).

The functions $\psi_{ii}^{(\alpha)}$ can be visualized as the components of a vector ψ in the space of the basis vectors for the irreducible representations of a group. The operator $\varepsilon_{ii}^{(\alpha)}$ projects the vector ψ in this direction. If there is not such component of ψ, then the result of operating with $\varepsilon_{ii}^{(\alpha)}$ upon ψ is of course zero. The geometric interpretation of the effect of $\varepsilon_{ik}^{(\alpha)}$ upon ψ is a little more complicated. It may be regarded as the projection of a previously rotated vector ψ onto the direction (αi). This rotation orients the component $\psi_{kk}^{(\alpha)}$ in the direction (αi). The action of $\varepsilon_{ik}^{(\alpha)}$ upon ψ therefore gives zero if there is no component of ψ in the direction (αk).

References

[1] I.G. Kaplan, *Symmetry of Many-Electron Systems*, Academic Press, New York, 1975.
[2] E. Wigner, *Group Theory*, Academic Press, New York, 1959.
[3] F.D. Murnagan, *The Theory of Group Representations*, John Hopkins Press, Baltimore, Maryland, 1938.
[4] M. Hamermesh, *Group Theory*, Addison-Wesley, Reading, Massachusetts, 1962.
[5] G.F. Koster, *Phys. Rev.* **109**, 227 (1958).

Appendix B

The Permutation Group

B.1 General Information

B.1.1 Operations with Permutation

Let us consider N arbitrary objects and enumerate these objects by the integers 1 to N. As is well known, we can form from N numbers $N!$ different permutations, which can be represented by the symbol

$$P = \begin{pmatrix} 1 & 2 & 3 & \ldots & N \\ i_1 & i_2 & i_3 & \ldots & i_N \end{pmatrix}, \tag{B.1}$$

where i_k stands for the number which, as a result of permutation, takes the place of the number k. The product of two permutations P_2P_1 is also a permutation, the effect of which is equivalent to the action of P_1 followed by that of P_2 and the product of three permutations is associative.

One can associate with every permutation (B.1) an inverse permutation

$$P^{-1} = \begin{pmatrix} i_1 & i_2 & i_3 & \ldots & i_N \\ 1 & 2 & 3 & \ldots & N \end{pmatrix}. \tag{B.2}$$

The Pauli Exclusion Principle: Origin, Verifications, and Applications, First Edition. Ilya G. Kaplan.

The effect of applying successively a permutation and its inverse is to leave the objects in their original positions, that is, this forms the identical permutation, which we denote by I.

The $N!$ permutations of N objects thus form a group which is called the *permutation group* or the *symmetric group*, and which we shall denote by π_N (this group is often denoted in literature by S_N). Each of $N!$ permutations is an independent element of the group π_N. But in applications they are usually considered as operators applied to functions of N arguments. Namely,

$$\begin{pmatrix} 1 & 2 & \dots & N \\ i_1 & i_2 & \dots & i_N \end{pmatrix} F(x_1, x_2, \dots, x_N) = F(x_{i_1}, x_{i_2}, \dots, x_{i_N}). \tag{B.3}$$

It is convenient to decompose the elements of a permutation group into products of *cycles*. A permutation is called a cycle if it can be written in the form

$$P_{i_1 i_2 \dots i_k} = \begin{pmatrix} i_1 & i_2 & i_3 & \dots & i_k & i_{k+1} & \dots & i_N \\ i_2 & i_3 & i_4 & \dots & i_1 & i_{k+1} & \dots & i_N \end{pmatrix}. \tag{B.4}$$

The cycle (B.4) is convenient to denote more shortly as

$$(i_1\, i_2 \dots i_k). \tag{B.4a}$$

The six elements of the group π_3 can be written as the cyclic permutations

$$I, P_{12}, P_{13}, P_{23}, P_{123}, P_{132}. \tag{B.5}$$

Every permutation can be represented in the form of a product of commuting cycles. For this purpose the numbers 1 and i_1 in (B.1) are taken to form the first two elements of a cycle. These are followed by the number that replaces i_1, the process being continued until the number that replaces 1 is reached. A similar procedure is followed for the remaining elements. For example,

$$\begin{pmatrix} 1 & 2 & 3 & 4 & 5 & 6 \\ 2 & 4 & 5 & 1 & 3 & 6 \end{pmatrix} = (124)(35)(6).$$

A cycle, by definition, is invariant under a cyclic permutation of its elements:

$$(i_1 i_2 i_3 \dots i_k) = (i_2 i_3 \dots i_k i_1) = (i_3 \dots i_k i_1 i_2) = \dots. \tag{B.6}$$

The number of elements in a cycle is called its *length*. It follows from the definition that if we raise a cycle to a power equal to its length, we obtain the identity permutation:

$$(i_1 i_2 \ldots i_k)^k = I. \tag{B.7}$$

From this we see that

$$(i_1 i_2 \ldots i_k)^{-1} = (i_1 i_2 \ldots i_k)^{k-1}. \tag{B.8}$$

A cycle of two elements is commonly known as a *transposition*. It is obvious that

$$(i_1 i_2) = (i_1 i_2)^{-1}. \tag{B.9}$$

The following rules are useful in manipulating permutations:

a. The product QPQ^{-1}, where Q and P are arbitrary permutations, is equal to the permutation that is obtained by letting the permutation Q act on P (in the sense of a permutation acting upon the arguments of a function). For example,

$$(13)(35)(13) = (15), \quad (123)(23)(123)^{-1} = (13).$$

b. The product of two permutations is independent of their order, if the cycles, which constitute the permutations, do not contain any common elements.
c. Two cycles which possess a common element can be combined by placing this element at the end of one cycle and at the beginning of the other, that is,

$$(ik \ldots lm)(mn \ldots q) = (ik \ldots lmn \ldots q). \tag{B.10}$$

For example,

$$(1245)(346) = (5124)(463) = (512463) \equiv (124635).$$

The product of two cycles belonging to the same permutation group is found by action of the first permutation on the second line of the second one written in the form (B.1). For example, let us consider a product of two cyclic permutations of the group π_5:

$$(13452)(12345) = (13452)\begin{pmatrix} 12345 \\ 23451 \end{pmatrix} = \begin{pmatrix} 12345 \\ 14523 \end{pmatrix} = (24)(35).$$

d. A cycle that results from multiplying other cycles may contain a number of identical elements. It is then useful to reduce this to cycles of the form

$$(ik\ldots lmi) \equiv (k\ldots lm) \tag{B.11}$$

Every cycle can always be written in the form of a product of transpositions. Such a representation, however, is not unique. Thus, for example,

$$\begin{aligned}(123\ldots k) &= (12)(23)\ldots(k-1\,k)\\ &= (1k)(1\,k-1)(1\,k-2)\ldots(12)\\ &= (21)(2k)(2k-1)\ldots(23) = \ldots.\end{aligned} \tag{B.12}$$

Nevertheless, the number of transpositions, which form a given permutation, always possesses a unique *parity*:

a permutation is classed as even or odd according to whether the number of transpositions, which it contains, is even or odd.

It is not difficult to show that any permutation can be represented as a product of transpositions of the form $(i-1, i)$, where $i-1$ and i are two consecutive numbers. Thus, for example,

$$(245) = (24)(45) = (23)(34)(23)(45).$$

This plays an important role in the determination of explicit forms for the representation matrices of the permutation group, see Section B.2.2.

B.1.2 Classes

As was shown, the $N!$ permutations of N objects satisfy the four group postulates and therefore form a group, which we denote by the symbol $\boldsymbol{\pi}_N$. The even permutations in $\boldsymbol{\pi}_N$ form a group by themselves known as the *alternating group*, and constitute a subgroup of $\boldsymbol{\pi}_N$. In addition to this, $\boldsymbol{\pi}_N$ possesses $N-1$ obvious further subgroups: $\boldsymbol{\pi}_{N-1}$, $\boldsymbol{\pi}_{N-2}$, …, and $\boldsymbol{\pi}_1$.

All permutations that are related to one another by the equation $P_i = QP_jQ^{-1}$, where Q is any permutation of $\boldsymbol{\pi}_N$, are, by definition, members of the same class. Since the permutation QP_jQ^{-1} is obtained by the action of Q upon P_j (see the previous section), the cyclic structures of P_i and P_j are identical, that is, the number and lengths of the cycles in the two permutations coincide. The permutations P_i and P_j differ only in the elements forming the cycles. Each class of $\boldsymbol{\pi}_N$ is therefore characterized by a particular partition of the N elements into cycles. The number of different classes is determined by the number of different ways of partitioning the number N into positive integral components, that is, is equal to the number of different integral solutions (with the exception of zero) of the equation

$$1\nu_1 + 2\nu_2 + \cdots + N\nu_N = N. \tag{B.13}$$

A set of numbers $\nu_1, \nu_2, \ldots, \nu_N$ which satisfies Eq. (B.13) uniquely defines a class of $\boldsymbol{\pi}_N$. We designate a class by the symbol $\{1^{\nu_1}2^{\nu_2}\ldots m^{\nu_m}\}$, where ν_k is the number of cycles of length k which appear in the permutations in the class. The class $\{1^N\}$ corresponds to the identity permutation. Thus, the six permutations of the group $\boldsymbol{\pi}_3$ are divided into three classes:

$$\{1^3\} : (1)(2)(3) \equiv I;$$
$$\{12\} : (1)(23), (2)(13), (3)(12);$$
$$\{3\} : (123), (132).$$

The group $\boldsymbol{\pi}_4$ contains five classes: $\{1^4\}$, $\{1^2 2\}$, $\{2^2\}$, $\{13\}$, and $\{4\}$.

We now derive a formula which connects the number of elements in a class with its cyclic structure. Let $\{1^{\nu_1}2^{\nu_2}\ldots m^{\nu_m}\}$ be an arbitrary class of the group $\boldsymbol{\pi}_N$. We place the N numbers in the cycles in their natural order, and let all the $N!$ permutations of the group operate upon this initial permutation, leaving the parentheses in place. Since the order of the cycles is unimportant, it is clear that $\nu_1 ! \, \nu_2 ! \ldots \nu_m !$, of the permutations will coincide. Cyclic permutations of the elements within a cycle also lead to identical permutations. For a single cycle of length k there are k such permutations, giving $2^{\nu_2}3^{\nu_3}\cdots m^{\nu_m}$ permutations in all. As a result, the order of the class is given by

$$g_{\{1^{\nu_1}2^{\nu_2}\ldots m^{\nu_m}\}} = \frac{N!}{\nu_1!\nu_2!2^{\nu_2}\ldots\nu_m!m^{\nu_m}}. \tag{B.14}$$

For example, the class $\{1^2 2\}$ of the group $\boldsymbol{\pi}_4$ contains $4!/(2!2) = 6$ elements, and the class $\{13\}$ contains $4!/3 = 8$ elements.

B.1.3 Young Diagrams and Irreducible Representations

Since the number of nonequivalent irreducible representations of a group is equal to the number of its classes, the nonequivalent irreducible representations of $\boldsymbol{\pi}_N$ are defined, as are the classes, by the different partitions of the number N into positive integral components. Each irreducible representation is typified by one such partition. The partitions are usually written in order of decreasing components $\lambda^{(i)}$:

$$\lambda^{(1)} + \lambda^{(2)} + \cdots + \lambda^{(m)} = N, \quad \lambda^{(1)} \geq \lambda^{(2)} \geq \cdots \geq \lambda^{(m)}. \tag{B.15}$$

Some of the $\lambda^{(i)}$ in (B.15) may coincide. It is clear that m cannot exceed N. Eq. (B.15) can be regarded as an equation for finding the possible partitions of N, and in this sense is entirely equivalent to Eq. (B.13). The partitions (B.15) can be depicted graphically by means of diagrams known as *Young diagrams*,[1] in which each number $\lambda^{(i)}$ is represented by a row of $\lambda^{(i)}$ cells. Young diagrams will be subsequently denoted by the symbol $[\lambda] \equiv \left[\lambda^{(1)}\lambda^{(2)}\ldots\lambda^{(m)}\right]$. The presence of several rows of identical length $\lambda^{(i)}$ will be indicated by a power of $\lambda^{(i)}$. For example,

$$[\lambda] = [2^2\, 1^2] =$$

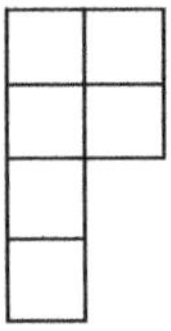

It is obvious that one can form from two cells only two Young diagrams:

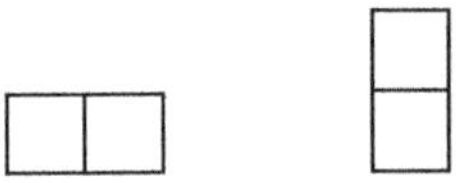

$$[2] \qquad [1^2] \tag{B.16}$$

For the group π_3 one can form from three cells three Young diagrams:

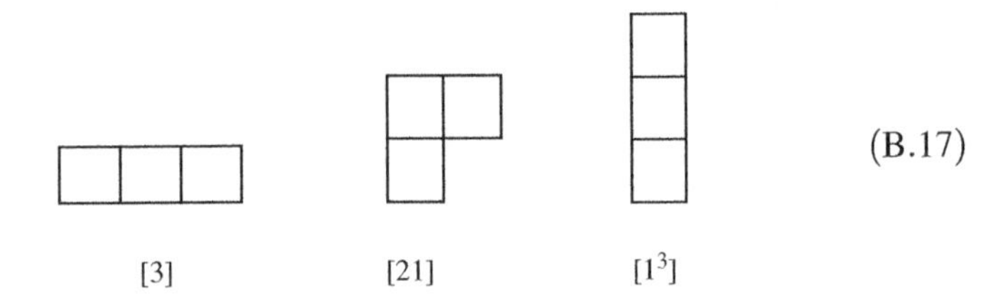

(B.17)

[1] Alfred Young (1873–1940) was a British mathematician. He elaborated his original approach to the permutation group in 1900–1935; the detailed account of Young's studies and bibliography can be found in the Rutherford book [1]. In the first two papers published in 1900 and 1902, Young introduces the fundamental concepts of, as they called after, Young diagrams and Young tableaux. It is interesting to note that the third paper was published in 1927, 25 years after the second. This 25 years gap will not be such surprising, if one takes into account that in 1908 Young was ordained and in 1910 became a country parish priest with his everyday duties. All his consequent papers on the permutation group were written when he was a clergyman. In 1934, Young being a parish priest was elected a Fellow of the Royal Society.

The group π_4 has five Young diagrams:

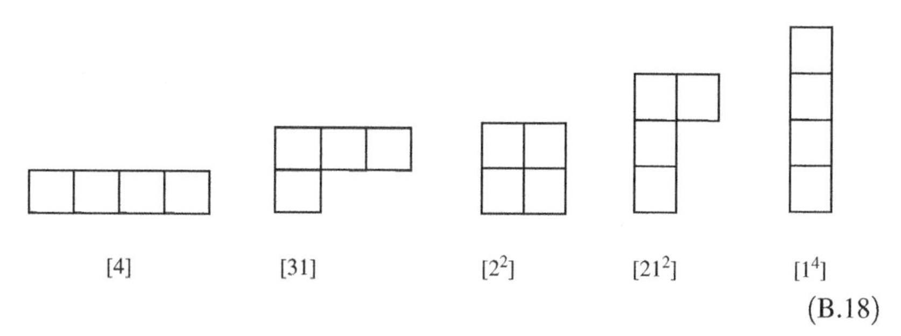

(B.18)

Each Young diagram uniquely corresponds to a specific irreducible representation of the group π_N. The irreducible representations of π_N are therefore usually enumerated by the symbol $[\lambda]$ for the corresponding Young diagrams, and are denoted by $\Gamma^{[\lambda]}$ or simply by $[\lambda]$. As will be shown, the assignment of a Young diagram determines the permutational symmetry of the basis functions for an irreducible representation, that is, determines the behavior of the basis functions under permutations of their arguments.

B.2 The Standard Young–Yamanouchi Orthogonal Representation

B.2.1 Young Tableaux

In general an irreducible representation $\Gamma^{[\lambda]}$ of the group π_N becomes reducible upon passing to the subgroup π_{N-1}, and can be decomposed into irreducible representations $\Gamma^{[\lambda']}$ of this subgroup (reduction in respect to a subgroup). The representations $\Gamma^{[\lambda']}$ into which $\Gamma^{[\lambda]}$ decomposes are determined by all the Young diagrams with $N-1$ cells that are obtained from the initial diagram upon removing one of its cells [1]. For example, the irreducible representation of the group π_5, characterized by the Young diagram $\Gamma^{[2^21]}$, splits up into the irreducible representations $\Gamma^{[2^2]}$ and $\Gamma^{[21^2]}$ of π_4:

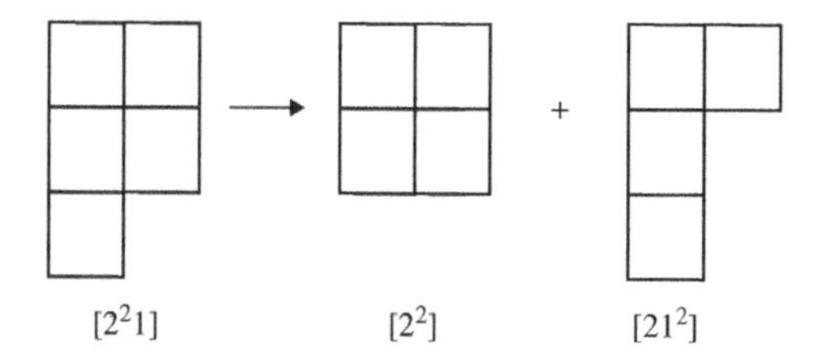

On passing from π_{N-1} to its subgroup π_{N-2}, the representation $\Gamma^{[\lambda']}$ in turn similarly split up into irreducible representations $\Gamma^{[\lambda'']}$ of π_{N-2}, the process continuing until the group π_1 is reached.

Linear combinations can be formed from the f_λ basis functions for an irreducible representation $\Gamma^{[\lambda]}$ such that in this new basis the matrices which correspond to the elements of the subgroups $\pi_{N-1}, \pi_{N-2}, \ldots$ (and which form representations of these subgroups) are all in reduced form. Basis functions of $\Gamma^{[\lambda]}$ chosen in this way simultaneously become basis functions for the representations $\Gamma^{[\lambda']}, \Gamma^{[\lambda'']}, \ldots$, on passing to the subgroups $\pi_{N-1}, \pi_{N-2}, \ldots$. Each basis function can be uniquely specified by the series of irreducible representations under which it transforms on passing from π_N to the various subgroups.

This choice of basis functions can be described graphically as follows. Each basis function is associated with a Young diagram $\Gamma^{[\lambda]}$ in the cells of which the integers are located ranging from one to N. The numbers are distributed among the cells in such a way that when the cell containing the number N is removed, one obtains the Young diagram $\Gamma^{[\lambda']}$ of the irreducible representation of π_{N-1} according to which the basis function transforms on passing to this subgroup. Subsequent removal of the cell containing $N-1$ gives the Young diagram $\Gamma^{[\lambda'']}$, etc. This process continues until we reach the diagram with just a single cell. It is clear that only those Young diagrams are allowed in which the number increases from left to right along the rows and down the columns (otherwise at some stage in the reduction forbidden Young diagrams will occur). We shall refer to a Young diagram of this kind of the number distribution as a *standard Young tableau*. The N numbers can be distributed among the cells of a Young diagram in $N!$ different ways, so in general there are $N!$ distinct Young tableaux. Henceforth, when speaking of Young tableaux we shall refer only to the standard tableaux, the number of which is always less than $N!$

Each basis function of an irreducible representation $\Gamma^{[\lambda]}$ can thus be associated with a Young tableau, and the dimension of the irreducible representation is given by the number of standard Young tableaux, that is, by the number of different ways of distributing the N numbers in a Young diagram such that they increase from left to right along the rows and down the columns.

For example, we can draw five standard Young tableaux for the representations $\Gamma^{[\lambda]} = \Gamma^{[2^3]}$ of the group π_6 (which we enumerate by the symbol $r^{(i)}$):

$$[\lambda] = [2^3]$$

$$r^{(1)}\ \begin{array}{|c|c|}\hline 1 & 2 \\ \hline 3 & 4 \\ \hline 5 & 6 \\ \hline \end{array} \quad r^{(2)}\ \begin{array}{|c|c|}\hline 1 & 2 \\ \hline 3 & 5 \\ \hline 4 & 6 \\ \hline \end{array} \quad r^{(3)}\ \begin{array}{|c|c|}\hline 1 & 3 \\ \hline 2 & 4 \\ \hline 5 & 6 \\ \hline \end{array} \quad r^{(4)}\ \begin{array}{|c|c|}\hline 1 & 3 \\ \hline 2 & 5 \\ \hline 4 & 6 \\ \hline \end{array} \quad r^{(5)}\ \begin{array}{|c|c|}\hline 1 & 4 \\ \hline 2 & 5 \\ \hline 3 & 6 \\ \hline \end{array} \tag{B.19}$$

The five basis functions which correspond to these Young tableaux are uniquely specified by the following series of representations of the groups π_5, π_4, π_3, and π_2:

$$\begin{aligned}
r^{(1)} &: [2^2 1] \to [2^2] \to [21] \to [2], \\
r^{(2)} &: [2^2 1] \to [21^2] \to [21] \to [2], \\
r^{(3)} &: [2^2 1] \to [2^2] \to [21] \to [1^2], \\
r^{(4)} &: [2^2 1] \to [21^2] \to [21] \to [1^2], \\
r^{(5)} &: [2^2 1] \to [21^2] \to [1^3] \to [1^2].
\end{aligned} \tag{B.20}$$

We arrange the Young tableaux in order of increasing deviation of the numbers from their natural order (as read along the rows, and from top to bottom), and we number the basis functions in the same order. We put first the Young tableaux in which the number two occurs in the first row, followed by those with two on the second row. Among the Young tableaux that have the number two on the same row, we place first those in which three occurs on the first row, followed by those with three on the second row, and so on. Succeeding numbers are dealt similarly.[2] The Young tableau in which the numbers are in their natural order is always first. We refer to this as the *fundamental tableau*. Since each basis function is specified by its Young tableau, one can use the tableaux to number the rows and columns of the irreducible representations according to which the basis functions transform.

If we are only interested in the dimension of an irreducible representation of a permutation group with N large, it is very inconvenient to write out all the possible Young tableaux. Instead, the dimension of an irreducible representation may be calculated by the following formula [1, 2]:

$$f_\lambda = \left\{ N! \prod_{i<j} (h_i - h_j) \right\} / (h_1!\, h_2! \cdots h_m!), \quad h_i = \lambda^{(i)} + m - i, \tag{B.21}$$

in which $\lambda^{(i)}$ is the length of row i and m is the number of rows in the Young diagram $[\lambda]$. For the representation $\Gamma^{[2^3]}$ formula (B.21) gives $f_\lambda = 5$, in agreement with the number of tableaux. So substituting the values of $\lambda^{(i)}$ and m into (B.21), we obtain $h_1 = 4$, $h_2 = 3$, and $h_3 = 2$, and hence $f_\lambda = (6!2)/(4!\ 3!\ 2!) = 5$.

[2] It is not difficult to satisfy oneself that the order of the Young tableaux in (B.19) obeys this rule.

B.2.2 Explicit Determination of the Matrices of the Standard Representation

The choice of basis functions for the irreducible representations of π_N which was explained in the previous section automatically gives an orthogonal set of functions. This is because the basis functions are each characterized by a distinct sequence of irreducible representations according to which they transform on passing from π_N to its subgroups. If the basis functions are also normalized, then in order to determine uniquely the matrices of the resulting orthogonal representation, it is only necessary to fix the phases of the matrix elements. For this purpose it suffices to give just the matrix of a transposition P_{i-1i}, since all the permutations in the group can be expressed in the form of a product of transpositions of this kind. Young (see Rutherford [1]) and Yamanouchi [3] showed that this representation may be chosen to be real and gave simple rules for constructing the matrices of P_{i-1i}. According to them, the matrix $\Gamma^{[\lambda]}(P_{i-1i})$ has the following nonzero elements:

1. the diagonal element $\Gamma^{[\lambda]}(P_{i-1i}) = 1$, if in tableau r the numbers i and $i-1$ occur on the same row;
2. the diagonal element $\Gamma^{[\lambda]}(P_{i-1i}) = -1$, if in tableau r the numbers i and $i-1$ occur on the same column;
3. $$\Gamma^{[\lambda]}(P_{i-1i}) = \begin{array}{c} \\ r \\ \\ t \\ \\ \end{array}\begin{bmatrix} & \vdots & & \vdots & \\ \cdots & -1/d & \cdots & [1-(1/d^2)]^{1/2} & \cdots \\ & \vdots & & \vdots & \\ \cdots & [1-(1/d^2)]^{1/2} & \cdots & 1/d & \cdots \\ & \vdots & & \vdots & \end{bmatrix},$$
   (columns labelled r and t)

   if the tableaux r and t differ only by a permutation of the number $i-1$ and i, and if the row in tableau r containing $i-1$ is above that containing i. The letter d denotes the axial distance between $i-1$ and i. This is defined as the number of vertical and horizontal steps that one must make in the Young tableau in order to move from $i-1$ to i. Thus, in tableau $r^{(4)}$ of (B.19), $d_{23} = 2$ and $d_{34} = 3$.

As an example, we write out the matrix of P_{23} of the irreducible representation $\Gamma^{[2^3]}$, the Young tableaux of which are given in (B.19) (vacant spaces denote zeros):

$$\Gamma^{[2^3]}(P_{23}) = \begin{array}{c} \\ r^{(1)} \\ r^{(2)} \\ r^{(3)} \\ r^{(4)} \\ r^{(5)} \end{array} \begin{array}{c} \begin{array}{ccccc} r^{(1)} & r^{(2)} & r^{(3)} & r^{(4)} & r^{(5)} \end{array} \\ \left[\begin{array}{ccccc} -\frac{1}{2} & & \frac{1}{2}\sqrt{3} & & \\ & -\frac{1}{2} & & \frac{1}{2}\sqrt{3} & \\ \frac{1}{2}\sqrt{3} & & \frac{1}{2} & & \\ & \frac{1}{2}\sqrt{3} & & \frac{1}{2} & \\ & & & & -1 \end{array} \right] \end{array}.$$

The representation which is defined by these rules is known as the *Young–Yamanouchi standard orthogonal representation*, or simply as the *standard representation*. Its matrix elements satisfy the following orthogonality relations:

$$\sum_r \Gamma^{[\lambda]}_{rt}(P)\Gamma^{[\lambda]}_{ru}(P) = \delta_{tu}, \quad \sum_r \Gamma^{[\lambda]}_{tr}(P)\Gamma^{[\lambda]}_{ur}(P) = \delta_{tu}. \tag{B.22}$$

These follow from the orthogonality of the representation, see Eq. (A.31).

$$\sum_P \Gamma^{[\lambda]}_{rs}(P)\Gamma^{[\bar{\lambda}]}_{ut}(P) = (N!/f_\lambda)\delta_{\lambda\bar{\lambda}}\delta_{ru}\delta_{st}, \tag{B.23}$$

$$\sum_{\lambda,r,t} (f_\lambda/N!)\Gamma^{[\lambda]}_{rt}(P)\Gamma^{[\lambda]}_{rt}(Q) = \delta_{PQ}. \tag{B.24}$$

These equations hold for any finite group; they form a particular case of Eqs. (A.38) and (A.39).

This method of constructing the standard representation of $\boldsymbol{\pi}_N$ ensures that the matrices are in fully reduced form with respect to the subgroups $\boldsymbol{\pi}_{N-1}, \boldsymbol{\pi}_{N-2}, \ldots, \boldsymbol{\pi}_2$. This means that if, for example, a permutation P_1 belongs to the group $\boldsymbol{\pi}_{N-2}$, then

$$\Gamma^{[\lambda]}_{rt}(P_1) = \delta_{\lambda_r''\lambda_t''}\Gamma^{[\lambda'']}_{r''t''}(P_1), \tag{B.25}$$

where r'' and t'' are the Young tableaux that are obtained from the tableaux r and t by removing the cells containing $N-1$ and N. If this process leads to different Young diagrams $[\lambda_r'']$ and $[\lambda_t'']$, then the matrix element (B.25) is equal to zero.

The matrices corresponding to permutations of the numbers $n_1+1, n_1+2, \ldots, N$ are diagonal in the Young tableaux for the first n_1 numbers. Denote these tableaux for the first n_1 numbers by the symbol r_1. Then the tableau r can be written in the form $r = (r_1\rho_2)$, where ρ_2 denotes the part of tableau r containing the numbers $n_1 + 1, \ldots, N$. Moreover, the matrix elements are equal for all the r_1 that arise from a particular Young diagram with n_1 cells, that is,

$$\Gamma^{[\lambda]}_{r_1\rho_2, \bar{r}_1\bar{\rho}_2}(P_2) = \delta_{r_1\bar{r}_1}\Gamma^{[\lambda]}_{r_1\rho_2, r_1\bar{\rho}_2}(P_2), \tag{B.26}$$

where P_2 denotes a permutation of the numbers $n_1 + 1, \ldots, N$. The matrices for the irreducible representations of groups $\boldsymbol{\pi}_3$–$\boldsymbol{\pi}_6$ are given in book [4], appendix 5. It can be easily seen that the representations given there satisfy all rules represented above.

As a simple example, we consider the standard irreducible representations of $\boldsymbol{\pi}_3$. This group has three irreducible representations, the Young diagrams of which have already been given, see Eq. (B.17).

The representations $\Gamma^{[3]}$ and $\Gamma^{[1^3]}$ are one-dimensional, since there is only one Young tableau for each of them:

1	2	3

1
2
3

According to the Young–Yamanouchi rules, the basis function for the $\Gamma^{[3]}$ representation does not change sign under any of the permutations, and generates the *symmetric* representation of $\boldsymbol{\pi}_3$. The characters of this representation are obviously all equal to unity. The basis function for the $\Gamma^{[1^3]}$ representation changes sign under odd permutations, and does not change sign under even permutations. A function such as this is said to be *antisymmetric*, and the one-dimensional representations generated by it is also said to be *antisymmetric*. The characters of this representation are equal to $(-1)^p$, where p is the parity of the permutation (p is the number of transposition forming the permutation P, we can put p equal to 0 for even permutations, and equal to 1 for odd permutations). In general, we note that the Young diagram $[N]$ for the group $\boldsymbol{\pi}_N$ always correspond to a one-dimensional symmetric representation, and that the diagram $[1^N]$ corresponds to a one-dimensional antisymmetric representation.

The $\Gamma^{[21]}$ representation possesses two standard Young tableaux:

$r^{(1)}$

1	2
3	

$r^{(2)}$

1	3
2	

$$d_{23} = 2 \tag{B.27}$$

The dimension of the representation is therefore equal to two. The Young tableaux (B.27) enumerate the rows and columns of the matrices of this representation.

$$[\lambda] = [21]$$

$$\begin{array}{ccc} I & P_{12} & P_{23} \\ \begin{bmatrix} 1 & \\ & 1 \end{bmatrix} & \begin{bmatrix} 1 & \\ & -1 \end{bmatrix} & \begin{bmatrix} -\frac{1}{2} & \frac{1}{2}\sqrt{3} \\ \frac{1}{2}\sqrt{3} & \frac{1}{2} \end{bmatrix} \\ P_{13} = P_{12}P_{23}P_{12} & P_{123} = P_{12}P_{23} & P_{123} = P_{123}^{-1} \\ \begin{bmatrix} \frac{1}{2} & -\frac{1}{2}\sqrt{3} \\ -\frac{1}{2}\sqrt{3} & \frac{1}{2} \end{bmatrix} & \begin{bmatrix} -\frac{1}{2} & \frac{1}{2}\sqrt{3} \\ -\frac{1}{2}\sqrt{3} & -\frac{1}{2} \end{bmatrix} & \begin{bmatrix} -\frac{1}{2} & -\frac{1}{2}\sqrt{3} \\ \frac{1}{2}\sqrt{3} & -\frac{1}{2} \end{bmatrix} \end{array}. \tag{B.28}$$

B.2.3 The Conjugate Representation[3]

We can associate with every standard representation $\Gamma^{[\lambda]}$ a conjugate (or *associated*) representation $\Gamma^{[\tilde{\lambda}]}$ of the same dimension, the matrices of which $\Gamma^{[\tilde{\lambda}]}(P)$ differ from those of the standard representation $\Gamma^{[\lambda]}(P)$ by a factor $(-1)^p$, where p is the parity of permutation.

The symmetric and antisymmetric representations, which were defined in the previous section, form the simplest example of conjugate representations. The first representation corresponds to the Young diagram $[N]$ and the second to the diagram $[1^N]$. The Young diagram $[1^N]$ is obtained from $[N]$ by changing the row into a column. In the general case it can be shown that for a representation characterized by the Young diagram $[\lambda]$, the conjugate representation is characterized by a diagram known as the *dual* Young diagram $[\tilde{\lambda}]$, which is obtained from $[\lambda]$ by interchanging its rows and columns. The rows and columns of the conjugate representation are enumerated by the Young tableaux of the Young diagram $[\tilde{\lambda}]$. To each Young tableau $r^{(i)}$, there corresponds a tableau $\tilde{r}^{(i)}$ which is obtained from $r^{(i)}$ by interchanging its rows and columns. Since the Young tableaux r are ordered according to the degree with which the numbers in it deviate from their natural order, the ordering of the tableaux $\tilde{r}$ is just the reverse. For example,

[3] Often it is referred to as the *adjoint* representation.

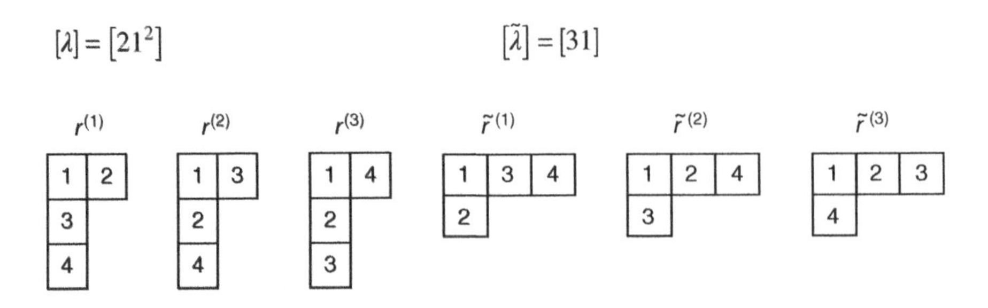

With the Young tableaux $\tilde{r}$ ordered in this way the rules for forming the matrices $\Gamma^{[\tilde{\lambda}]}(P_{i-1\,i})$ are, except for the off-diagonal elements $\Gamma^{[\tilde{\lambda}]}_{\tilde{r}\tilde{t}}(P_{i-1\,i})$, the same as those for the $\Gamma^{[\lambda]}(P_{i-1\,i})$. For two tableaux $\tilde{r}$ and $\tilde{t}$ that differ only by a permutation of $i-1$ and i the off-diagonal elements are given by

$$\Gamma^{[\tilde{\lambda}]}_{\tilde{r}\tilde{t}}(P_{i-1\,i}) = -\left[1-\left(\frac{1}{d^2}\right)\right]^{1/2}. \tag{B.29}$$

The minus sign in front of the root is introduced in order to satisfy the condition

$$\Gamma^{[\tilde{\lambda}]}(P_{i-1\,i}) = -\Gamma^{[\lambda]}(P_{i-1\,i}).$$

Thus, the representation that is conjugate to $[\lambda] = [21]$ (see Eq. B.28) is characterized by the same Young diagram [21]. The ordering of the Young tableaux $\tilde{r}$, however, is different:

$\tilde{r}^{(1)}$

1	3
2	

$\tilde{r}^{(2)}$

1	2
3	

By the rules for constructing the transposition matrices in the standard representation, we obtain, taking (B.29) into account,

$$\Gamma^{[\tilde{21}]}(P_{12}) = \begin{bmatrix} -1 & \\ & 1 \end{bmatrix}, \quad \Gamma^{[\tilde{21}]}(P_{23}) = \begin{bmatrix} \frac{1}{2} & -\frac{1}{2}\sqrt{3} \\ -\frac{1}{2}\sqrt{3} & -\frac{1}{2} \end{bmatrix}. \tag{B.30}$$

We see that these matrices differ by a factor -1 from the corresponding matrices in the standard representation $\Gamma^{[21]}$ (see Eq. B.28). The remaining matrices in the $\Gamma^{[\tilde{21}]}$ representation are obtained by multiplying the matrices (B.30) together. Consequently, the $\Gamma^{[\tilde{21}]}$ matrices will differ by a factor $(-1)^p$ from those in the $\Gamma^{[21]}$ representation.

B.2.4 *The Construction of an Antisymmetric Function from the Basis Functions for Two Conjugate Representations*

Consider the direct product $\Gamma^{[\lambda_1]} \times \Gamma^{[\lambda_2]}$ of two irreducible representations of $\boldsymbol{\pi}_N$. This can be resolved into its irreducible components by a suitable linear transformation of the basis functions. We determine the condition that the antisymmetric representation occurs in this decomposition. The characters of the antisymmetric representation are equal to $(-1)^p$. Substituting this into formula (A.46) and making use of the orthogonality relations (A.41), we find that

$$\begin{aligned} a^{[1^N]} &= \left(\frac{1}{N!}\right)\sum_P (-1)^p \chi^{[\lambda_1]}(P)\chi^{[\lambda_2]}(P) \\ &= \left(\frac{1}{N!}\right)\sum_P \chi^{[\tilde{\lambda}_1]}(P)\chi^{[\lambda_2]}(P) = \delta_{\tilde{\lambda}_1\lambda_2}. \end{aligned} \tag{B.31}$$

The antisymmetric representation therefore occurs once only in the reduction of $\Gamma^{[\lambda_1]} \times \Gamma^{[\lambda_2]}$, and only if $[\lambda_2] = [\tilde{\lambda}_1]$. In other words, it is impossible to form an antisymmetric linear combination from products of the basis functions $\psi_{r_1}^{[\lambda_1]}\psi_{r_2}^{[\lambda_2]}$, unless the representations $\Gamma^{[\lambda_1]}$ and $\Gamma^{[\lambda_2]}$ are conjugate to one another. If $\Gamma^{[\lambda_1]}$ and $\Gamma^{[\lambda_2]}$ are conjugate, then one, and only one, antisymmetric function can be formed in this way. It is easy to verify that this antisymmetric function is given by

$$\psi^{[1^N]} = \frac{1}{\sqrt{f_\lambda}}\sum_r \psi_r^{[\lambda]}\psi_{\tilde{r}}^{[\tilde{\lambda}]}, \tag{B.32}$$

in which the summation is taken over all the f_λ Young tableaux, and the $1/(f_\lambda)^{1/2}$ factor is a normalization constant. Let an arbitrary permutation P operates on the function (B.32). Then

$$\begin{aligned} P\psi^{[1^N]} &= \left[\frac{1}{(f_\lambda)^{1/2}}\right]\sum_r \sum_{t,\tilde{u}} \Gamma_{tr}^{[\lambda]}(P)\Gamma_{\tilde{u}\tilde{r}}^{[\tilde{\lambda}]}(P)\psi_t^{[\lambda]}\psi_{\tilde{u}}^{[\tilde{\lambda}]} \\ &= \left[\frac{1}{(f_\lambda)^{1/2}}\right]\sum_{t,\tilde{u}}\left[\sum_r \Gamma_{tr}^{[\lambda]}(P)\Gamma_{ur}^{[\lambda]}(P)\right](-1)^p\psi_t^{[\lambda]}\psi_{\tilde{u}}^{[\tilde{\lambda}]}. \end{aligned}$$

The summation over r in this equation gives δ_{tu}, in accordance with Eq. (B.22). As a result, we obtain

$$P\psi^{[1^N]} = (-1)^p\psi^{[1^N]},$$

which demonstrates the correctness of Eq. (B.32).

B.2.5 Young Operators

In Appendix A it was shown that one can obtain a set of basis functions for an irreducible representation of any finite group by applying the operators (A.75) to some arbitrary function. We shall refer to these operators, when specialize to the standard orthogonal Young–Yamanouchi representation, as *Young operators*, and denote them by $\omega_{rt}^{[\lambda]}$. *The normalized Young operator* is of the form

$$\omega_{rt}^{[\lambda]} = \left(\frac{f_\lambda}{N!}\right)^{1/2} \sum_P \Gamma_{rt}^{[\lambda]}(P)P. \tag{B.33}$$

The summation over P runs over all the $N!$ permutations of the group π_N. The operator $\omega_{rt}^{[\lambda]}$ differs from the $\varepsilon_{ik}^{(\alpha)}$ by the absence of the sign for complex conjugation on the matrix elements (the matrices in a representation $\Gamma^{[\lambda]}$ are all real), and also in the factor in front of the summation. This factor is chosen so that application of the operator $\omega_{rt}^{[\lambda]}$ to a product of N orthogonal functions $\phi_a(i)$ produces a normalized function (see below).

One can form f_λ^2 operators $\omega_{rt}^{[\lambda]}$ for each representation $\Gamma^{[\lambda]}$, and since

$$\sum_\lambda f_\lambda^2 = N!,$$

there are altogether $N!$ such operators for the group π_N. According to Eqs. (B.23) and (B.24), the quantities $(f_\lambda/N!)^{1/2}\Gamma_{rt}^{[\lambda]}(P)$ form an orthogonal matrix of dimension $N!$. Eq. (B.33) may therefore be regarded as an orthogonal transformation from $N!$ permutation operators P to $N!$ operators $\omega_{rt}^{[\lambda]}$. The inverse transformation is given by

$$P = \sum_{\lambda,r,t} \left(\frac{f_\lambda}{N!}\right)^{1/2} \Gamma_{rt}^{[\lambda]}(P)\omega_{rt}^{[\lambda]}. \tag{B.34}$$

From Eq. (A.76) we can obtain an expression for the product of two Young operators, if we substitute in it $\varepsilon_{rt}^{[\lambda]} = (f_\lambda/N!)^{1/2}\omega_{rt}^{[\lambda]}(P)$. As a result, we obtain

$$\omega_{rt}^{[\lambda]}\omega_{us}^{[\tilde{\lambda}]} = \left(\frac{N!}{f_\lambda}\right)^{1/2} \delta_{\lambda\tilde{\lambda}}\delta_{tu}\omega_{rs}^{[\lambda]}. \tag{B.35}$$

Upon representing P in the form of a sum (B.34) and using (B.35), we find that

$$P\omega_{rt}^{[\lambda]} = \sum_u \Gamma_{ur}^{[\lambda]}(P)\omega_{ut}^{[\lambda]}. \tag{B.36}$$

that is, only the operators with the same second index transform into each other under permutations. It follows from (B.36) that a set of f_λ Young operators $\omega_{rt}^{[\lambda]}$ with a fixed second index form a basis for a representation $\Gamma^{[\lambda]}$. The Young operators $\omega_{rt}^{[\lambda]}$ can therefore be regarded as basis vectors for an irreducible representation $\Gamma^{[\lambda]}$.

If the order of the operators on the left-hand side of (B.36) is reversed, we obtain a different result. Using Eqs. (B.34) and (B.35), we find that

$$\omega_{rt}^{[\lambda]} P = \sum_u \Gamma_{tu}^{[\lambda]}(P)\omega_{ru}^{[\lambda]}. \tag{B.37}$$

We consider several examples of the Young operators. The group $\boldsymbol{\pi}_2$ possesses two one-dimensional representations, with a single Young operator corresponding to each of them:

$$\omega^{[2]} = \frac{1}{\sqrt{2}}(I + P_{12}), \tag{B.38}$$

$$\omega^{[1^2]} = \frac{1}{\sqrt{2}}(I - P_{12}). \tag{B.39}$$

Operator (B.38) is a symmetrizing operator and (B.39) is an antisymmetrizing operator. The Young operators for the one-dimensional representations $\Gamma^{[3]}$ and $\Gamma^{[1^3]}$ of $\boldsymbol{\pi}_3$ are similarly constructed:

$$\omega^{[3]} = \left(\frac{1}{\sqrt{6}}\right)(I + P_{12} + P_{23} + P_{13} + P_{123} + P_{132}), \tag{B.40}$$

$$\omega^{[1^3]} = \left(\frac{1}{\sqrt{6}}\right)(I - P_{12} - P_{23} - P_{13} + P_{123} + P_{132}). \tag{B.41}$$

The representation $\Gamma^{[21]}$ is two-dimensional. Four independent Young operators can therefore be constructed in this case. Taking the matrix elements of the standard representation from Eq. (B.28), we obtain from formula (B.33) the following four operators:

$$\begin{aligned}
\omega_{11}^{[21]} &= \left(\frac{1}{\sqrt{12}}\right)(2I + 2P_{12} - P_{23} - P_{13} - P_{123} - P_{132}),\\
\omega_{12}^{[21]} &= \left(\frac{1}{2}\right)(P_{23} - P_{13} + P_{123} - P_{132}),\\
\omega_{21}^{[21]} &= \left(\frac{1}{2}\right)(P_{23} - P_{13} - P_{123} + P_{132}),\\
\omega_{22}^{[21]} &= \left(\frac{1}{\sqrt{12}}\right)(2I - 2P_{12} + P_{23} + P_{13} - P_{123} - P_{132}).
\end{aligned} \tag{B.42}$$

B.2.6 *The Construction of Basis Functions for the Standard Representation from a Product of* N *Orthogonal Functions*

We consider a set of N orthonormal functions $\phi_a(i)$ (the letter i in the argument of the function stands for the set of variables upon which ϕ_a depends):

$$\int \phi_a^* \phi_b dV = \delta_{ab}. \tag{B.43}$$

We form the function Φ_0, a product of N functions ϕ_a:

$$\Phi_0 = \phi_1(1)\phi_2(2)\ldots\phi_N(N). \tag{B.44}$$

From Φ_0 we can produce $N!$ different functions in two ways: by permuting the arguments, leaving the functions in place, or by permuting the functions (it is the same, as to permute the function labels), leaving the order of the arguments unchanged. We distinguish those permutations that permute the functions by placing a bar over the symbols for them. It is not difficult to check that there is a simple relationship between the two sorts of permutations, namely:

$$\bar{P}\Phi_0 = P^{-1}\Phi_0. \tag{B.45}$$

The permutation P and $\bar{P}$ act upon different objects, and hence commute with one another:

$$\bar{P}Q\Phi_0 = Q\bar{P}\Phi_0. \tag{B.46}$$

The set of permutations P and $\bar{P}$ constitutes two commuting groups $\boldsymbol{\pi}_N$ and $\bar{\boldsymbol{\pi}}_N$.

Let us permute the arguments of the function Φ_0. We obtain $N!$ mutually orthogonal functions:

$$\Phi_P = P\Phi_0 = \phi_1(i_1)\phi_2(i_2)\ldots\phi_N(i_N). \tag{B.47}$$

The orthogonality of the functions Φ_P follows from the condition (B.43). The set of the functions Φ_P forms a basis for the regular representation of $\boldsymbol{\pi}_N$ (see Section A.2.9).

Linear combinations of the Φ_P functions, which form a basis for the standard orthogonal representations, are obtained by applying the Young operator (B.33) to the function Φ_0:

$$\Phi_{rt}^{[\lambda]} = \omega_{rt}^{[\lambda]}\Phi_0 = \left(\frac{f_\lambda}{N!}\right)^{1/2} \sum_P \Gamma_{rt}^{[\lambda]}(P)\Phi_P. \tag{B.48}$$

In accordance with (B.36)

$$P\Phi_{rt}^{[\lambda]} = \sum_u \Gamma_{ur}^{[\lambda]}(P)\Phi_{ut}^{[\lambda]}. \tag{B.49}$$

Functions $\Phi_{rt}^{[\lambda]}$ with a fixed second index therefore transform into each other under permutations of the arguments in conformity with the general rule, Eq. (A.74). The Young tableau r that corresponds to the first index enumerates the basis functions for the representation $\Gamma^{[\lambda]}$. This tableau thus characterizes the symmetry of the function $\Phi_{rt}^{[\lambda]}$ under permutations of the arguments. The Young tableau t characterizes the symmetry of $\Phi_{rt}^{[\lambda]}$ under permutations of the functions ϕ_a. In order to prove this last statement, we apply a permutation $\bar{P}$ to the function (B.48). By virtue of Eqs. (B.46) and (B.45),

$$\bar{P}\Phi_{rt}^{[\lambda]} = \omega_{rt}^{[\lambda]}\bar{P}\Phi_0 = \omega_{rt}^{[\lambda]}P^{-1}\Phi_0. \tag{B.50}$$

Expressing the operators on the right-hand side of (B.50) in the form (B.37) we finally obtain

$$\bar{P}\Phi_{rt}^{[\lambda]} = \sum_u \Gamma_{ut}^{[\lambda]}(P)\Phi_{ru}^{[\lambda]}. \tag{B.51}$$

Functions $\Phi_{rt}^{[\lambda]}$ with a fixed first index therefore transform into each other under permutations of the ϕ_a functions (or the function labels).

In this way one can form from the $N!$ functions (B.47) $N!$ linearly independent functions (B.48), which we denote by $\Phi_{rt}^{[\lambda]}$. If each function $\Phi_{rt}^{[\lambda]}$ is represented by a point on a graph, then the $N!$ points fall into squares, each of which is characterized by a particular Young diagram $[\lambda]$, and contains f_λ^2 points. The rows of the squares are numbered by the Young tableaux r and the columns by the tableaux t. The functions corresponding to points in a single column of a square transform into each other under permutations of the argument, and functions that correspond to points on a single row of a square transform into each other under permutations of the ϕ_a.

For example, for $N=4$, the $4!$ functions $\Phi_{rt}^{[\lambda]}$ fall into five squares, corresponding to the five Young diagrams (B.18),

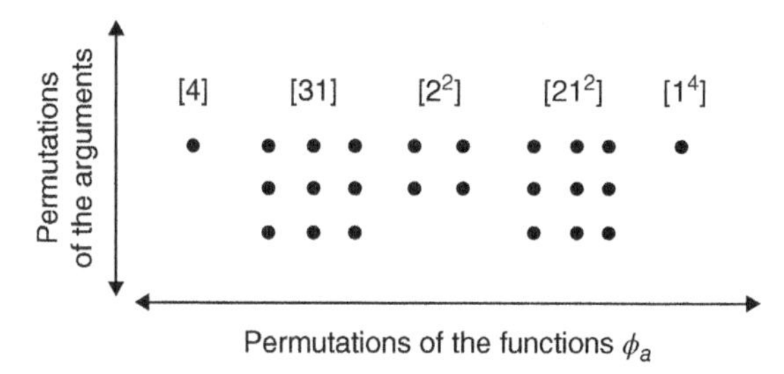

As was shown in Appendix A for any finite group, the basis functions that transform according to different columns of an irreducible representation are orthogonal

to one another. However, the basis functions that transform according to a same column but belong to different bases are no longer orthogonal to each other, see Eq. (A.78). The $\Phi_{rt}^{[\lambda]}$ functions, however, satisfy a stronger orthogonality condition than this equation, and are orthogonal with respect to all their indices; that is,

$$\int \Phi_{rt}^{[\lambda]^*} \Phi_{us}^{[\bar{\lambda}]} dV = \delta_{\lambda\bar{\lambda}} \delta_{ru} \delta_{ts}. \tag{B.52}$$

The additional orthogonality, as compared to Eq. (A.78), in the second indices follows from the fact that for fixed first indices the functions $\Phi_{rt}^{[\lambda]}$ and $\Phi_{rs}^{[\lambda]}$ belong to a single basis for the group $\bar{\pi}_N$.

Eq. (B.52) can be derived directly by using expression (B.48) for $\Phi_{rt}^{[\lambda]}$ and applying the orthogonality of Φ_P and the orthogonality relations (B.23). Thus,

$$\int \Phi_{rt}^{[\lambda]^*} \Phi_{us}^{[\bar{\lambda}]} dV = \frac{(f_\lambda f_{\bar{\lambda}})^{1/2}}{N!} \sum_{P,Q} \Gamma_{rt}^{[\lambda]}(P) \Gamma_{us}^{[\bar{\lambda}]}(Q) \int \Phi_P^* \Phi_Q dV$$

$$= \frac{(f_\lambda f_{\bar{\lambda}})^{1/2}}{N!} \sum_{P} \Gamma_{rt}^{[\lambda]}(P) \Gamma_{us}^{[\bar{\lambda}]}(P) = \delta_{\lambda\bar{\lambda}} \delta_{ru} \delta_{ts}$$

Let us assume that m of the functions in Eq. (B.44) are identical; that is,

$$\Phi_0 = \phi_1(1)\ldots\phi_1(m)\phi_2(m+1)\ldots\phi_{N-m+1}(N). \tag{B.53}$$

The number of independent functions $\Phi_{rt}^{[\lambda]}$, which may be formed by applying the f_λ^2 Young operators $\omega_{rt}^{[\lambda]}$ to the function (B.53), is less than f_λ^2, since some of the $\Phi_{rt}^{[\lambda]}$ functions are now linearly dependent and some reduce to zero. This is because the function (B.53) is symmetric with respect to permutations of the arguments among the m function ϕ_1. Only those Young tableaux in which the first m numbers all appear on the same line are allowed. The possible Young diagrams $[\lambda]$ therefore all embody a diagram consisting of a single row with m cells. Since the total number of cells in a Young diagram is equal to N, no columns in the diagram may have more than $N-m+1$ cells.

For example, in the case

$$\Phi_0 = \phi_1(1)\phi_1(2)\phi_1(3)\phi_2(4)\phi_3(5)$$

the following Young diagrams are allowed for the group π_5:

$$[\lambda] = [5],\ [41],\ [32],\ [31^2].$$

The Young diagrams $[2^2 1]$, $[21^3]$, and $[1^5]$ are not realized.

This result does not depend upon the factorization of Φ_0 into a product of one-electron functions ϕ_a, and can be put into the following more general form:

the types of permutation symmetry that can be realized for a function of N *variables,* m *of which are identical and the remaining* N − m *distinct, are those for which the Young diagrams have one row with not less than m cells, and no column with more than* N − m + 1 cells.

The *nonstandard* representations of the permutation group were first introduced by the author in publication [5] and described in detail in the book [4].

References

[1] D.E. Rutherford, *Substitutional Analysis*, Hafner Publishing Co., New York/London, 1968.
[2] F.D. Murnagan, *The Theory of Group Representations*, John Hopkins Press, Baltimore, Maryland, 1938.
[3] T. Yamanouchi, *Proc. Phys. Math. Soc. Jpn.* **19**, 436 (1937).
[4] I.G. Kaplan, *Symmetry of Many-Electron Systems*, Academic Press, New York, 1975.
[5] I.G. Kaplan, *Sov. Phys. - JETP* **14**, 568 (1962).

Appendix C

The Interconnection between Linear Groups and Permutation Groups

C.1 Continuous Groups

C.1.1 Definition

In Appendices A and B we considered *discrete groups*, that is, groups whose elements form a discrete set. A discrete set of elements can always be enumerated by positive integers. However, there is a large class of groups whose elements form a *continuous set*. Each element is characterized by a number of parameters which can vary continuously. Such groups are known as *continuous groups*. I present here a short description of the main features of continuous groups, for more detailed account see rigorous courses, for example, Pontriagin's book [1].

For example, the set of transformations

$$x' = a_1 x + a_2, \tag{C.1}$$

in which the parameters a_1 and a_2 can assume any value on the number scale axis (from $-\infty$ to $+\infty$), forms a two-parameter continuous group. Any variation of the parameter, even if infinitely small, leads to a new element of the group.

The Pauli Exclusion Principle: Origin, Verifications, and Applications, First Edition. Ilya G. Kaplan.

We can also formally associate a value of some parameter a with every element G_a of a discrete group. The parameter corresponding to the product of two elements, $G_c = G_a G_b$, is equal to c. The discrete function

$$c = \psi(a,b), \tag{C.2}$$

which places the value of the parameter of the product c in correspondence with the values of the parameters a and b, determines the multiplication table for the group.

All the elements of a discrete group can be described by specifying a single parameter, which can assume g discrete values. In the case of a continuous group the function (C.2) becomes a continuous function of its arguments, the number of parameters, which characterize an element of the group, being arbitrary and even infinite.

In what follows we consider groups of linear transformations in an n-dimensional vector space, the transformation being characterized by a finite number of parameters. These groups form a particular case of *Lie groups*. The group of transformations

$$x_i' = f_i(x_1,\ldots,x_n;a_1,\ldots,a_r) \quad i = 1,2,\ldots n, \tag{C.3}$$

in which the functions are analytic in the parameters a_ρ, is known as a *r-parameter Lie group.*

The application of two such transformations in succession,

$$x_i' = f_i(x_1,\ldots,x_n;a_1,\ldots,a_r)$$

and

$$x_i'' = f_i\left(x_1',\ldots,x_n';b_1,\ldots,b_r\right),$$

is equivalent to a third transformation,

$$x_i''' = f_i(x_1,\ldots,x_n;c_1,\ldots,c_r),$$

in which the parameters are functions of the parameters of the first two transformations:

$$c_k = \psi_k(a_1,\ldots,a_r;b_1,\ldots,b_r) \quad k = 1,2,\ldots r. \tag{C.4}$$

The transformations (C.3) also have to satisfy the remaining three group postulates.

Thus, in the case of two-parameter group presented in (C.1), Eq. (C.4) are of the form

$$c_1 = b_1 a_1, \quad c_2 = b_1 a_2 + b_2. \tag{C.5}$$

The unit element in this group is characterized by the values of the parameters $a_1 = 1$ and $a_2 = 0$. According to the definition of an inverse element, the successive application of an element and its inverse must give the identity transformation ($c_1 = 1$ and $c_2 = 0$). Hence we find from (C.5) the following relation between the parameters of a transformation and its inverse:

$$\overline{a}_1 = \frac{1}{a_1}, \quad \overline{a}_2 = -\frac{a_2}{a_1}$$

The basic concepts and theorems, which were established in Appendix A for discrete groups, can immediately be generalized to continuous groups. Only those statements that are based upon the finiteness of the order of the group cease to have any meaning (e.g., the statement that the dimensions of the irreducible representations are divisors of the order of the group). The concepts of subgroups, classes of conjugate elements, the reducibility of a representation, and so on, are carried over unchanged to continuous groups. Since there is a continuous set of elements in a group, each irreducible representation consists of a continuous set of matrices. The number of nonequivalent irreducible representations is infinite, though these form a discrete series, in which the dimensions of all the irreducible representations are finite, that is, the number of basis functions which transform into each other under the operations of the group is finite.[1]

For continuous groups a summation over the elements of a group is replaced by an integration over the domain over which the parameter can vary. The volume element in the integration is chosen so that the integral of an arbitrary continuous function of the group parameter is invariant under any transformation of the group. This is known as *invariant integration*. The mathematical requirement that an integral over the group parameters be invariant is written as

$$\int_R f(R) d\tau_R = \int_R f(RQ) d\tau_R, \tag{C.6}$$

in which the group operators R and Q are functions of the parameters, and the integration is taken over the whole domain over which the parameters of the group

[1] This is true only for groups known as *compact* continuous groups [1], to which practically all the continuous groups that are used in physical applications belong (the Lorentz group is an exception). The condition that the integral (C.6) be invariant cannot be met for noncompact groups, since the integration over the domain of the parameters diverges.

can vary. This equation is the natural extension to a continuous group of relation (A.7) which was proved in Appendix A.

The replacement of the summations in (A.38) and (A.41) by invariant integrations leads to the following form for the orthogonality relations for continuous groups:

$$\int_R \Gamma_{ik}^{(\alpha)}(R)^* \Gamma_{lm}^{(\beta)}(R) d\tau_R = \delta_{\alpha\beta}\delta_{il}\delta_{km}\left(\frac{1}{f_\alpha}\right)\int_R d\tau_R, \tag{C.7}$$

$$\int_R \chi^{(\alpha)}(R)^* \chi^{(\beta)}(R) d\tau_R = \delta_{\alpha\beta}\int_R d\tau_R. \tag{C.7a}$$

Formula (A.46) now becomes

$$a^{(\alpha)} = \left(\frac{1}{\displaystyle\int_R d\tau_R}\right)\int_R \chi^{(\Gamma)}(R)\chi^{(\alpha)}(R)^* d\tau_R. \tag{C.8}$$

C.1.2 Examples of Linear Groups

Let us consider a vector space of n dimensions. We subject the space to a linear transformation, the matrix of which is $\{a_{ik}\}$. Every vector in the space $\mathbf{x}$ is transformed into a new vector $\mathbf{x}'$ with components

$$x_i' = \sum_{k=1}^{n} a_{ik}x_k. \tag{C.9}$$

A linear transformation is said to be *nonsingular* if the determinant of the transformation matrix is nonzero. The set of all nonsingular linear transformations in an n-dimensional space forms a group known as the *general linear group*, which is denoted by $\mathbf{GL}_n$. The product of two linear transformations is, of course, also a linear transformation, the matrix of which is found by multiplying the matrices of the individual transformations. Since the determinant of a transformation is nonzero, there exists for every transformation A an inverse transformation A^{-1}, the successive application of these two transformations leads to the identity transformation. Finally, the associative rule is satisfied.

Since each linear transformation is determined by its matrix, these matrices form an n-dimensional representation of the group $\mathbf{GL}_n$. Generally speaking, the transformation matrices are complex, and therefore in order to characterize a transformation, $2n^2$ real parameters must be specified.

If we restrict the linear transformation to unitary transformations, we obtain the *group of unitary transformations* in an n-dimensional space, which is denoted by the symbol $\mathbf{U}_n$. The elements of the transformation matrices satisfy the unitary conditions

$$\sum_l a_{il}^* a_{kl} = \delta_{ik}. \tag{C.10}$$

This gives n^2 equations among the $2n^2$ parameters,[2] and so the number of independent parameters is equal to n^2. Every element of the unitary group is therefore characterized by specifying n^2 real parameters.

Unitary transformations with determinant equal to unity form the group of unitary unimodular transformations, which is denoted by $\mathbf{SU}_n$ (the symbol is derived from the other name for this group: *special unitary group*). In addition to satisfying relations (C.10), the elements of the transformation matrices of $\mathbf{SU}_n$ must satisfy the condition that the determinant of a transformation is equal to unity. Every transformation in the group $\mathbf{SU}_n$ is therefore characterized by specifying $n^2 - 1$ real parameters.

The set of all real unitary transformations, that is, of all orthogonal transformations, forms a subgroup of the unitary group which is called the *orthogonal group* $\mathbf{O}_n$. The orthogonality conditions (see Eq. (A.31)) give $n + n(n-1)/2$ equations among the n^2 parameters of a matrix. The elements of the orthogonal group are therefore characterized by specifying $n(n-1)/2$ parameters. Since the determinant of a matrix is equal to that of its transpose, it follows from Eq. (A.30) that the square of the determinant of an orthogonal transformation is equal to 1, and so the determinant itself can only assume two values, +1 or −1. An orthogonal transformation with determinant equal to +1 corresponds to a *rotation of the space* about the origin of coordinates. An orthogonal transformation with determinant equal to −1 constitutes a combination of a rotation and an *inversion of the space* with respect to the origin of coordinates.

If the orthogonal transformations are restricted to those whose determinant is equal to plus one, we obtain the *rotation group in n-dimensional space* $\mathbf{R}_n$. For $n = 3$, $\mathbf{R}_n$ becomes the rotation group in three-dimensional space $\mathbf{R}_3$, which is widely used in physics.

Every subgroup of the orthogonal group is called a *point group*. The rotation group $\mathbf{R}_3$ is a continuous point group. Finite point groups, the elements of which consist of combinations of rotations through definite angles and of reflections in planes, are used in the theory of molecules and crystals. The various point groups are classified in many books on quantum mechanics and group theory, for example, see Ref. [2], sections 3.8 and 3.9.

[2] For $i = k$, there are n equations for the moduli of the matrix elements; for $i \neq k$, there are $n(n-1)/2$ equations for the real parts of the matrix elements and as many again for the imaginary parts.

If the groups of linear transformations described above are placed in the order, in which they are contained in each other, we obtain the following chain:

$$\mathbf{GL}_n \supset \mathbf{U}_n \supset \left\{ \begin{array}{c} \mathbf{O}_n \\ \mathbf{SU}_n \end{array} \right\} \supset \mathbf{R}_n.$$

C.1.3 Infinitesimal Operators

Any finite transformation of a continuous group can be represented as a succession of infinitely small transformations. The theory of Lie groups, see Pontriagin [1] or Racah [3], shows that a finite transformation is uniquely determined by specifying the infinitely small transformation. All the matrices of an irreducible representation, moreover, can be expressed in terms of the matrices which correspond to infinitely small transformations in the same representations. In short, a continuous group is completely characterized by its infinitely small transformations.

Let us rewrite Eq. (C.3) in a more compact form, denoting the set of r parameters a_ρ by the letter a and the set of n coordinates x_i by the letter x:

$$x'_i = f_i(x;a). \tag{C.3a}$$

It is possible to write another equivalent expression for the x'_i, which represents the identity operation:

$$x'_i = f_i(x';0). \tag{C.3b}$$

Instead of differentiating (C.3a), an infinitely small change in the x'_i can be obtained by introducing an infinitely small parameter in (C.3b):

$$x'_i + dx'_i = f_i(x';\delta a). \tag{C.11}$$

Expanding the right-hand side of (C.11) in a McLaurin series and confining ourselves to terms of the first order of smallness, we obtain

$$dx'_i = \sum_{\rho=1}^{r} \left[\left(\frac{\partial f_i(x',a)}{\partial a_\rho} \right) \right]_{a=0} \delta a_\rho. \tag{C.12}$$

We denote

$$\left[\left(\frac{\partial f_i(x,a)}{\partial a_\rho} \right) \right]_{a=0} = u_{i\rho}(x). \tag{C.13}$$

Then, since

$$x_i + dx_i = x_i + \sum_{\rho=1}^{r} u_{i\rho}(x)\delta a_\rho, \tag{C.14}$$

the specification of the r vectors (C.13) in an n-dimensional space defines an infinitely small change in the position of any point in this space. Let us determine the change in an arbitrary function $F(x)$ for an infinitely small change in x:

$$\begin{aligned} dF(x) &= \sum_{i=1}^{n} \frac{\partial F}{\partial x_i} dx_i = \sum_{i=1}^{n} \frac{\partial F}{\partial x_i} \sum_{\rho=1}^{r} u_{i\rho}(x)\delta a_\rho \\ &= \sum_{\rho=1}^{r} \delta a_\rho \left(\sum_{i=1}^{n} u_{i\rho}(x) \frac{\partial}{\partial x_i} \right) F = \sum_{\rho=1}^{r} \delta a_\rho I_\rho F, \end{aligned} \tag{C.15}$$

in which I_ρ denotes the operator

$$I_\rho = \sum_{i=1}^{n} u_{i\rho}(x) \left(\frac{\partial}{\partial x_i} \right) \tag{C.16}$$

The operators I_ρ are called the *infinitesimal operators* of the group. Their number is equal to the number of group parameters. According to Eq. (C.15), an infinitely small change in the function $F(x)$ is produced by the action of a linear combination of infinitesimal operators upon the function. It can be shown [3] that the commutator of two infinitesimal operators can always be expressed as a linear combination

$$I_\rho I_\sigma - I_\sigma I_\rho \equiv [I_\rho, I_\sigma] = \sum_{\tau=1}^{r} c_{\rho\sigma}^{\tau} I_\tau. \tag{C.17}$$

The coefficients $c_{\rho\sigma}^{\tau}$ are known as *the structure constants of the group*.

As an example, we consider the group of rotations in three-dimensional space. According to the classification in the previous section, the group $\mathbf{R}_3$ is the group of orthonormal transformations in which the determinants of all the transformations are equal to plus one. The number of independent parameters is equal to $n(n-1)/2 = 3$. We may take as those parameters either the Euler angles or the coordinates of a vector directed along the axis of rotation and whose length is equal to the angle of rotation. It is clear that the specification of such a vector uniquely defines a rotation.

We denote the Cartesian coordinates of a point in the three-dimensional space by x, y, and z, and consider a transformation in which one of the parameters is varied, for instance, by a rotation through an angle α_x about the x-axis. In this case, Eq. (C.3) assume the following form:

$$x' = x,$$
$$y' = y\cos\alpha_x - z\sin\alpha_x, \tag{C.18}$$
$$z' = y\sin\alpha_x + z\cos\alpha_x.$$

From formula (C.13) we find[3]

$$u_{xx} = 0, \quad u_{yx} = -z, \quad u_{zx} = y. \tag{C.19}$$

Consequently,

$$I_x = y\left(\frac{\partial}{\partial z}\right) - z\left(\frac{\partial}{\partial y}\right). \tag{C.20}$$

The two other infinitesimal operators are obtained in a similar manner:

$$I_y = z\left(\frac{\partial}{\partial x}\right) - x\left(\frac{\partial}{\partial z}\right), \quad I_z = x\left(\frac{\partial}{\partial y}\right) - y\left(\frac{\partial}{\partial x}\right). \tag{C.21}$$

A direct calculation shows that the operators (C.20) and (C.21) satisfy the following commutation relations:

$$[I_x, I_y] = -I_z, \quad [I_y, I_z] = -I_x, \quad [I_z, I_x] = -I_y. \tag{C.22}$$

C.2 The Three-Dimensional Rotation Group

C.2.1 Rotation Operators and Angular Momentum Operators

In the preceding section it was shown that the change in a function due to an infinitely small rotation is determined by the infinitesimal operators (C.20) and (C.21). Thus, for an infinitely small rotation about an axis x_ρ through an angle $\delta\alpha$,

$$dF(x) = (\delta\alpha) I_\rho F(x). \tag{C.23}$$

[3] In (C.19) and subsequent equations the same notation will be used for both the coordinates and the parameters of a transformation.

In quantum mechanics the change in a wave function due to infinitely small rotations is expressed in terms of the angular momentum operator **J** [4]. The operator **J** is a vector quantity and its three components J_ρ are related to the infinitesimal operators I_ρ by the simple equation,[4]

$$J_\rho = -iI_\rho. \tag{C.24}$$

The commutation relations for the J_ρ follow from (C.22):

$$[J_x, J_y] = iJ_z, \quad [J_y, J_z] = iJ_x, \quad [J_z, J_x] = iJ_y. \tag{C.25}$$

In general, angular momentum operators **J** are defined in quantum mechanics as operators whose components obey the commutation relations (C.25). Spin angular momentum operators are also covered by this definition.

We now show that the angular momentum operators determine the change in a function of the coordinates due to an arbitrary finite rotation. According to Eqs. (C.23) and (C.24), an arbitrary function $F(x)$ receives an increment

$$dF(x) = i(\delta\alpha)J_\rho F(x), \tag{C.26}$$

due to an infinitely small rotation about the x_ρ axis through an angle $\delta\alpha$. The function $F(x)$ becomes

$$F'(x) = \left(1 + i\delta\alpha J_\rho\right)F(x). \tag{C.27}$$

A rotation through a finite angle α can be carried out by k consecutive rotations through an angle α/k. If k is a sufficiently large number, Eq. (C.27) holds for each rotation. Hence for a rotation through an angle α, the function $F(x)$ becomes

$$F'(x) = \lim_{k\to\infty} \left[1 + i(\alpha/k)J_\rho\right]^k F(x) = e^{i\alpha J_\rho} F(x). \tag{C.28}$$

The obvious generalization of this equation to a rotation about an axis in the direction of an arbitrary unit vector **n** is

$$F'(x) = [\exp i\alpha(\mathbf{n}\cdot\mathbf{J})]F(x). \tag{C.29}$$

[4] The eigenvalues of J_ρ are expressed in units of $\hbar = h/2\pi$, where h is Planck's constant. The effect of multiplying I_ρ by i is to make the operator Hermitian, so that the matrices of J_ρ satisfy the Hermitian condition $\tilde{J}^*_\rho = J_\rho$.

From this equation it follows that

$$R_{\mathbf{n},\alpha} = \exp i\alpha(\mathbf{n}\cdot\mathbf{J}), \tag{C.30}$$

can be regarded as an operator for a finite rotation about an axis **n** through an angle α. In order to find the effect of (C.30) upon a function of the coordinates, one must write the operator in the form of a power series and determine the effect of each term of the series upon the function.

A rotation through an angle α about an axis **n** can be transformed into a rotation through the same angle about some other axis $\mathbf{n}'$. For this purpose the axis **n** is brought into coincidence with $\mathbf{n}'$ by means of a rotation R. The rotation through the angle α is then carried out, and finally **n** is returned to its original position by the rotation R^{-1}, that is,

$$R_{\mathbf{n},\alpha} = R^{-1} R_{\mathbf{n}',\alpha} R. \tag{C.31}$$

All rotations through a given angle therefore belong to the same class of the group $\mathbf{R}_3$. In order to specify a particular rotation operation it is necessary to give its axis and angle of rotation, while the class of $\mathbf{R}_3$, to which the rotation belongs, is characterized only by the magnitude of the angle of rotation.

C.2.2 Irreducible Representations[5]

It can be shown that knowledge of the commutation relation for the infinitesimal operators of a continuous group is sufficient to determine all the irreducible representations of the group. The infinitesimal operators of the three-dimensional rotation group are just the angular momentum operators whose commutation relations are given by Eq. (C.25). In order to find the possible irreducible representations, we make use of the set of eigenfunctions of one of the operators J_ρ, for example, the eigenfunctions of J_z. We impose upon these eigenfunctions the requirement that they be simultaneously eigenfunctions of the operator for the square of the angular momentum.

$$J^2 = J_x^2 + J_y^2 + J_z^2. \tag{C.32}$$

This requirement can always be satisfied, since the operators J_z and J^2 commute.[6] The eigenfunctions $\psi_m^{(j)}$ are therefore characterized by two indices the values of which determine the eigenvalues of J^2 and J_z. In addition, these indices, by means

[5] For detailed treatment of the representations of the three-dimensional rotation group, see Refs. [5–7].
[6] We recall that the necessary and sufficient condition for two operators to possess a common set of eigenfunctions is that the operators commute [4].

of the commutation relations (C.25), determine the effect of applying J_x, J_y, and J^2 to the functions $\psi_m^{(j)}$. Thus, one finds, see Refs. [4, 5],

$$\begin{aligned} J^2\psi_m^{(j)} &= j(j+1)\psi_m^{(j)}, \\ J_z\psi_m^{(j)} &= m\psi_m^{(j)}, \\ (J_x + iJ_y)\psi_m^{(j)} &= [(j-m)(j+m+1)]^{1/2}\,\psi_{m+1}^{(j)}, \\ (J_x - iJ_y)\psi_m^{(j)} &= [(j+m)(j-m+1)]^{1/2}\,\psi_{m-1}^{(j)}, \end{aligned} \tag{C.33}$$

where j can assume integer and half-integer values only, and, for given j, m can assume the $2j+1$ values within the bounds $|m| \leq j$:

$$m = j,\ j-1, \ldots, -j. \tag{C.34}$$

It follows from Eqs. (C.33) that functions $\psi_m^{(j)}$ with fixed j transform into one another under infinitely small rotations. Since, according to Eq. (C.30), the operator for a finite rotation can be expressed in terms of angular momentum operators, the $\psi_m^{(j)}$ also transforms among themselves under finite rotations, that is,

$$R_{\mathbf{n},\alpha}\psi_m^{(j)} = \sum_{m'} D_{m'm}^{(j)}(R_{\mathbf{n},\alpha})\psi_{m'}^{(j)}, \tag{C.35}$$

where the coefficients $D_{m'm}^{(j)}(R_{\mathbf{n},\alpha})$ form a matrix which corresponds to a rotation through an angle α about an axis $\mathbf{n}$. The $2j+1$ functions $\psi_m^{(j)}$ therefore form a $(2j+1)$-dimensional basis for a representation of the rotation group $\mathbf{R}_3$. It can be shown that this representation, which is usually denoted by $D^{(j)}$, is irreducible [5]. Its matrix elements satisfy orthogonality relations which form a particular case of relations (C.7). The volume element for $\mathbf{R}_3$ can be written as $d\tau_R = \sin\beta d\beta d\alpha d\gamma$, where α, β, and γ are the Euler angles [5]. As a result, we obtain, instead of (C.7),

$$\int D_{m\mu}^{(j)}(R)^* D_{m'\mu'}^{(j')}(R) d\tau_R = \left[\frac{8\pi^2}{2j+1}\right]\delta_{jj'}\delta_{mm'}\delta_{\mu\mu'}. \tag{C.36}$$

An explicit form for the representation matrices $D^{(j)}(\alpha, \beta, \gamma)$, expressed in terms of the Euler angles has been obtained by Wigner [6].

It is not difficult to find the characters of the class which corresponds to a rotation through an angle α. Since the classes of $\mathbf{R}_3$ are typified by the angle of rotation only and are independent of the direction of the axis of rotation, we take the z-axis to be axis of rotation. The rotation operator (C.30) therefore assumes the form

$$R_{z,\alpha} = e^{i\alpha J_z}. \tag{C.37}$$

In order to find the effect of this operator on a function $\psi_m^{(j)}$, we expand the exponent and note that

$$J_z^k \psi_m^{(j)} = m^k \psi_m^{(j)},$$

that is, in every term in the series the operator is replaced by its eigenvalue. We therefore obtain

$$R_{z,\alpha}\psi_m^{(j)} = e^{im\alpha}\psi_m^{(j)}, \tag{C.38}$$

The matrix of the rotation $R_{z,\alpha}$ is thus diagonal and its character in the representation $D^{(j)}$ is given by

$$\chi^{(j)}(\alpha) = \sum_{m=-j}^{j} e^{im\alpha} = \frac{e^{i(j+1)\alpha} - e^{-ij\alpha}}{e^{i\alpha} - 1}. \tag{C.39}$$

Upon dividing the numerator and denominator of (C.39) by $e^{i\alpha/2}$, we obtain an expression which is more convenient in computations,

$$\chi^{(j)}(\alpha) = \frac{\sin(j + 1/2)\alpha}{\sin(\alpha/2)}. \tag{C.40}$$

From this formula it follows that

$$\chi^{(j)}(-\alpha) = \chi^{(j)}(\alpha).$$

Two rotations through identical angles but in opposite directions thus belong to the same class; that is, a class is characterized by the absolute value of the angle of rotation.

The set of rotations about a fixed axis forms the two-dimensional rotation group $\mathbf{R}_2$. This group is Abelian, and all its irreducible representations are one-dimensional. It follows from Eq. (C.38) that the $2j+1$ basis functions for an irreducible representation of $D^{(j)}$ of the group $\mathbf{R}_3$ belong to different irreducible representations of $\mathbf{R}_2$, each of which is characterized by a particular value of the number m.

According to Eq. (C.38), when j and, consequently, m are half-integers, a basis function $\psi_m^{(j)}$ changes sign on rotation by 2π. The identity transformations correspond to a rotation by 4π. As a result, two matrices $D^{(j)}(R_{n,\alpha})$ which differ from one another in sign correspond to each rotation $R_{n,\alpha}$. In order to obtain well-defined representations, Bethe [8] introduced the group known as the *double group* of three-dimensional rotations, in which the rotation by 2π is regarded as an element that is distinct from the identity. Rotations by $\alpha(\alpha \le 2\pi)$ and $2\pi + \alpha$ are regarded as different elements, as a result of which the volume of the double group is twice that of the ordinary group $\mathbf{R}_3$.

When j is integral ($j = l$), the explicit form of the basis functions can be found by solving the first two equations of (C.33) directly. However, it is first of all necessary to replace J^2 and J_z by their expressions in spherical polar coordinates. The solutions of the resulting equations are the spherical harmonics $Y_{lm}(\theta, \phi)$, which are very well known in mathematical physics. It can be shown [5] that the matrix elements of the representation $D^{(j)}$ for integral $j = l$ are proportional to the spherical harmonics:

$$D^{(l)}_{m'0}(\alpha,\beta,0) = \left(\frac{4\pi}{2l+1}\right)^{1/2} Y_{lm'}(\beta,\alpha),$$

$$D^{(l)}_{0m}(0,\beta,\gamma) = (-1)^m \left(\frac{4\pi}{2l+1}\right)^{1/2} Y_{lm}(\beta,\gamma),$$

where α, β, and γ are the Euler angles. The matrix elements $D^{(j)}_{m'm}(\alpha,\beta,\gamma)$ are therefore known as *generalized spherical harmonics* of the jth order.

The irreducible representations of the three-dimensional rotation group are thus characterized by a number j which can be either integral or half-integral. The dimension of an irreducible representation $D^{(j)}$ is $2j + 1$. The basis functions are characterized by two indices, the index j denoting the particular irreducible representation $D^{(j)}$ and the index m enumerating the basis functions within the representation. On the other hand, the functions $\psi^{(j)}_m$ are eigenfunctions of the operators for the square of the angular momentum and the projection of it upon the z-axis. The indices j and m thus have an additional interpretation: index j determines the eigenvalue of the square of the angular momentum, which is equal to $j(j+1)$, and m the projection of the angular momentum upon the z-axis.[7]

C.2.3 *Reduction of the Direct Product of Two Irreducible Representations*

Let us construct the direct product of two irreducible representations $D^{(j_1)} \times D^{(j_2)}$. This representation has the dimension $(2j_1 + 1)(2j_2 + 1)$ and is reducible. Products

[7] One usually says that the magnitude of the angular momentum vector $\mathbf{J}$ is j, meaning that the maximum value of the projection of the angular momentum is equal to j.

of the basis functions for the individual representations $\psi^{j_1}_{m_1}\psi^{j_2}_{m_2}$, form a basis for this representation. Since each of the functions $\psi^{j_i}_{m_i}$ is an eigenfunction of the operator J_{iz}, it is easy to see that a product of two such functions is an eigenfunction of the operator

$$J_z = J_{1z} + J_{2z}, \tag{C.41}$$

with eigenvalue

$$m = m_1 + m_2. \tag{C.42}$$

However, the products $\psi^{j_1}_{m_1}\psi^{j_2}_{m_2}$ are not eigenfunctions of the operator $J^2 = (\boldsymbol{J}_1 + \boldsymbol{J}_2)^2$. By forming linear combinations that are eigenfunctions of this operator, we carry out the reduction of the direct product into irreducible representations $D^{(j)}$. In order to find which irreducible representations $D^{(j)}$ can occur in the decomposition of this representation, we write out, using rule (C.42), the $(2j_1+1)(2j_2+1)$ eigenvalues m of the projection of the angular momentum. These eigenvalues are then divided into sets of $2j+1$ values of the form (C.34), each set corresponding to an irreducible representation $D^{(j)}$. We obtain as a result the desired decomposition[8]

$$D^{(j_1)} \times D^{(j_2)} \doteq D^{(j_1+j_2)} + D^{(j_1+j_2-1)} + \dots D^{(|j_1-j_2|)}. \tag{C.43}$$

Each irreducible representation $D^{(j)}$ occurs once in the decomposition, the values of j lying in the interval

$$|j_1 - j_2| \le j \le j_1 + j_2, \qquad j_1 + j_2 + j \ \text{ is integral.} \tag{C.44}$$

If one constructs a triangle with an integral perimeter length and sides of lengths j_1, j_2, and j, the condition, which the sides must obey, is just Eq. (C.44). Eq. (C.44) is called the *triangle rule with integral perimeter*, and is denoted by $\Delta(j_1 j_2 j)$.

The linear transformation from the basis functions of the reducible representation to basis functions for the irreducible representations which occur in the decomposition (C.43) is carried out by means of the matrix of the Clebsch–Gordan coefficients (see Appendix A, Section A.2.8),

$$\psi^{(j)}_m = \sum_{m_1, m_2} \psi^{j_1}_{m_1}\psi^{j_2}_{m_2} \langle j_1 m_1, j_2 m_2 | jm \rangle. \tag{C.45}$$

[8] The decomposition of the direct product into irreducible components, Eq. (C.43), is equivalent to finding the possible values of j that the total angular momentum vector can assume when the angular momentum vectors of the individual wave functions are coupled together. The decomposition (C.43) is therefore known in quantum mechanics as the rule for coupling angular momenta.

The Clebsch–Gordan coefficients $\langle j_1m_1, j_2m_2|jm\rangle$ are nonzero only if conditions (C.42) and (C.44) are fulfilled. The summation over m_2 in Eq. (C.45) is therefore purely formal, since $m_2 = m - m_1$. According to Eq. (A.62) the $\langle j_1m_1, j_2m_2|jm\rangle$ coefficients satisfy the orthogonality relations

$$\sum_{j,m} \langle j_1m_1, j_2m_2|jm\rangle\langle j_1m_1', j_2m_2'|jm\rangle = \delta_{m_1m_1'}\delta_{m_2m_2'}$$
$$\sum_{m_1,m_2} \langle j_1m_1, j_2m_2|jm\rangle\langle j_1m_1, j_2m_2|j'm'\rangle = \delta_{jj'}\delta_{mm'}. \tag{C.46}$$

A number of different explicit expressions for the Clebsch–Gordan coefficients in terms of the parameters j_1, j_2, j, m_1, and m_2 which occur in them may be found in books of Edmonds [9] and Yutsis et al. [10]. References to available tables can also be found in these books. From the explicit expressions for the Clebsch–Gordan coefficients we obtain the following relations:

$$\langle j_1m_1, j_2m_2|jm\rangle = \begin{cases} (-1)^{j_1+j_2-j}\langle j_1-m_1, j_2-m_2|j-m\rangle, & \text{(C.47a)} \\ (-1)^{j_1+j_2-j}\langle j_2m_2, j_1m_1|jm\rangle, & \text{(C.47b)} \\ (-1)^{j_1-m_1}\left(\dfrac{2j+1}{2j_2+1}\right)^{\frac{1}{2}}\langle j_1m_1, j-m|j_2-m_2\rangle. & \text{(C.47c)} \end{cases}$$

If two identical angular momenta are coupled to give a resultant angular momentum of zero, the Clebsch–Gordan coefficients differ only in phase and are given by

$$\langle jm, j-m|00\rangle = (-1)^{j-m}(2j+1)^{-1/2}. \tag{C.48}$$

We note that when constructing a function of the form (C.45) the ordering of the functions being coupled is important. If the angular momenta are equal, $j_1 = j_2 = j$, then according to Eq. (C.47b), the function (C.45) becomes multiplied by a phase factor when the positions of $\psi^j_{m_1}$ and $\psi^j_{m_2}$ are permuted,

$$P_{12}\psi^{(J)}_M[j(1)j(2)] = \psi^{(J)}_M[j(2)j(1)] = (-1)^{2j-J}\psi^{(J)}_M[j(1)j(2)]. \tag{C.49}$$

Instead of the Clebsch–Gordan coefficients, one often makes use of more symmetric coefficients known as *3-j symbols*. The 3-j symbols are written in the form of two-rowed matrices,

$$\begin{pmatrix} j_1 & j_2 & j_3 \\ m_1 & m_2 & m_3 \end{pmatrix}.$$

They are related to the Clebsch–Gordan coefficients by the equation

$$\langle j_1 m_1, j_2 m_2 | jm \rangle = (-1)^{-j_1+j_2-m}(2j+1)^{1/2} \begin{pmatrix} j_1 & j_2 & j \\ m_1 & m_2 & -m \end{pmatrix}. \tag{C.50}$$

The 3-j symbols are invariant under any even permutation of their columns, and are multiplied by $(-1)^{j_1+j_2+j_3}$ under any odd permutation of their columns.

C.2.4 Reduction of the Direct Product of k Irreducible Representations. $3n-j$ Symbols

The decomposition of the direct product of k irreducible representations

$$D^{(j_1)} \times D^{(j_2)} \times \cdots \times D^{(j_k)} \tag{C.51}$$

into irreducible components is carried out by successively reducing the direct products of pairs of irreducible representations. A reduction such as this can be produced in several different ways. In cases where a precise specification is unnecessary, we shall denote the reduction scheme by a capital letter such as A, B,.... Thus for $k=3$ there are two different reduction schemes:

(a) The direct product of the representations $D^{(j_1)} \times D^{(j_2)}$ is reduced first, followed by the reduction of the direct product of the representations that are obtained in this way, $D^{(j_{12})}$, with the remaining representation $D^{(j_3)}$.
(b) The direct product of the last two representations $D^{(j_2)} \times D^{(j_3)}$ is reduced first, following which the direct products of $D^{(j_1)}$ with the representations obtained in the first reduction, $D^{(j_{23})}$, are reduced.

The order of reduction is conveniently denoted by enclosing the reducible representations in brackets. The two reduction schemes discussed above can then be written in the form

$$\begin{aligned} &(\mathrm{a}) \quad \left(D^{(j_1)} \times D^{(j_2)}\right) \times D^{(j_3)}, \\ &(\mathrm{b}) \quad D^{(j_1)} \times \left(D^{(j_2)} \times D^{(j_3)}\right). \end{aligned} \tag{C.52}$$

The matrix which reduces the direct product (C.51) is a generalization of the matrix of Clebsch–Gordan coefficients. Since the reduction process consists of successively reducing pairs of representations, the coefficients in the linear transformation from the basis functions for the initial, reducible representation to basis functions $\psi_m^{(j)}$ for the resulting, irreducible representations can be expressed as products of the corresponding Clebsch–Gordan coefficients. The reduction process involves $k-2$ intermediate irreducible representations $D^{(j_{\text{int}})}$ and therefore the basis functions $\psi_m^{(j)}$ are characterized by $k-2$ additional quantum numbers j_{int}. These quantum numbers can be regarded as the eigenvalues of $k-2$ intermediate angular momentum operators $\boldsymbol{J}_{\text{int}}$ which occur in the vector coupling of k angular momenta $\mathbf{J}_1, \mathbf{J}_2, \ldots, \mathbf{J}_k$. The representations $D^{(j)}$ which are obtained on reducing the direct product (C.51) are independent of the reduction scheme. The form of the basis functions $\psi_m^{(j)}$, however, does depend upon the reduction scheme, since the set of Clebsch–Gordan coefficients, which expresses the $\psi_m^{(j)}$ in terms of the basis functions of the direct product (C.51), depends upon the particular reduction scheme. The basis functions, which are obtained from different schemes, are usually referred to in quantum mechanics as wave functions for states with different *coupling schemes for the angular momenta*. We denote the coupling scheme for the angular momenta by capital letters A, B,..., the same as for reduction scheme. A basis function for a representation $D^{(j)}$ with scheme A for coupling the angular momenta is written

$$\psi_m^{(j)}\left((j_1 \ldots j_k)(j_{\text{int}})^A\right), \tag{C.53}$$

in which $(j_{\text{int}})^A$ denotes the set of $k-2$ intermediate angular momenta which occur in the given scheme for coupling the angular momenta $\mathbf{J}_1, \mathbf{J}_2, \ldots, \mathbf{J}_k$.

Two sets of basis functions for a representation $D^{(j)}$ which differ in the coupling scheme for the angular momenta can be transformed into each other by means of some orthogonal transformation,

$$\psi_m^{(j)}\left((j_1 \ldots j_k)(j_{\text{int}})^A\right) = \sum_{(j_{\text{int}})^B} \psi_m^{(j)}\left((j_1 \ldots j_k)(j_{\text{int}})^B\right)\left\langle (j_{\text{int}})^B | (j_{\text{int}})^A \right\rangle^{(j)}. \tag{C.54}$$

The coefficients in this transformation cannot depend on the index of the basis function, that is, upon m, and due to the orthogonality of the basis functions, are given by the following integral in configuration space:

$$\left\langle (j_{\text{int}})^B | (j_{\text{int}})^A \right\rangle^{(j)} = \int \psi_m^{(j)}\left((j_1 \ldots j_k)(j_{\text{int}})^B\right)^* \psi_m^{(j)}\left((j_1 \ldots j_k)(j_{\text{int}})^A\right) dV. \tag{C.55}$$

We shall refer to the matrix for the transformation (C.54) as the *transformation matrix for the three-dimensional rotation group*. This matrix is obviously diagonal in j. We now seek an explicit form of it for the case of three coupled angular momenta.

There are two possible schemes for coupling three angular momenta, corresponding to the two reduction schemes (C.52). The basis functions, which correspond to these coupling schemes, are found by the successive application of formula (C.45):

$$\psi_m^{(j)}((j_1j_2)j_{12}j_3) = \sum_{m_{12},m_3} \psi_{m_{12}}^{(j_{12})}\psi_{m_3}^{(j_3)}\langle j_{12}m_{12}, j_3m_3|jm\rangle$$
$$= \sum_{\substack{m_1,m_2,\\ m_3,m_{12}}} \psi_{m_1}^{(j_1)}\psi_{m_2}^{(j_2)}\psi_{m_3}^{(j_3)}\langle j_1m_1, j_2m_2|j_{12}m_{12}\rangle\langle j_{12}m_{12}, j_3m_3|jm\rangle. \tag{C.56}$$

Similarly,

$$\psi_m^{(j)}(j_1(j_2j_3)j_{23}) = \sum_{\substack{m_1,m_2,\\ m_3,m_{23}}} \psi_{m_1}^{(j_1)}\psi_{m_2}^{(j_2)}\psi_{m_3}^{(j_3)}\langle j_1m_1, j_{23}m_{23}|jm\rangle\langle j_2m_2, j_3m_3|j_{23}m_3\rangle. \tag{C.57}$$

We find the transformation matrix, which connects the basis functions (C.56) and (C.57), from formula (C.55). Making use of the orthonormality of the $\psi_{m_i}^{(j_i)}$ and of the reality of the Clebsch–Gordan coefficients, we obtain

$$\langle (j_1j_2)j_{12}j_3|j_1(j_2j_3)j_{23}\rangle^{(j)} = \sum_{\substack{m_1,m_2,m_3\\ m_{12},m_{23}}} \langle j_1m_1, j_2m_2|j_{12}m_{12}\rangle \times \langle j_{12}m_{12}, j_3m_3|jm\rangle\langle j_2m_2, j_3m_3|j_{23}m_{23}\rangle \times \langle j_1m_1, j_{23}m_{23}|jm\rangle. \tag{C.58}$$

The matrix elements (C.58) do not depend upon the m_i, and are functions of six variables. They are related to coefficients known as *Racah W coefficients*[9] by the following equation:

$$\langle (j_1j_2)j_{12}j_3|j_1(j_2j_3)j_{23}\rangle^{(j)} = [(2j_{12}+1)(2j_{23}+1)]^{1/2}W(j_1j_2jj_3;j_{12}j_{23}). \tag{C.59}$$

In a similar way it can be shown that when the order in which the angular momenta are coupled is changed, the elements of the transformation matrix can be expressed in terms of Racah coefficients. Thus, if we interchange angular momenta j_2 and j_3, we have [12]

[9] These coefficients were introduced by Racah in his classic papers on the theory of complex atomic spectra, see Ref. [11].

$$\langle (j_1 j_2) j_{12} j_3 | (j_1 j_3) j_{13} j_2 \rangle^{(j)} = [(2j_{12}+1)(2j_{13}+1)]^{\frac{1}{2}} W(j_{12} j_3 j_2 j_{13}; j j_1). \qquad \text{(C.60)}$$

From (C.58) it follows that the $W(j_1 j_2 j j_3; j_{12} j_{23})$ coefficients are nonzero only if the four triangle conditions

$$\Delta(j_1 j_2 j_{12}), \quad \Delta(j_2 j_3 j_{23}), \quad \Delta(j_{12} j_3 j), \quad \text{and} \quad \Delta(j_1 j_{23} j)$$

are satisfied.

Instead of the Racah coefficients, one often uses more symmetric expressions known as 6-j symbols which are written in the form of a two-rowed matrix,

$$\begin{Bmatrix} j_1 & j_2 & j_{12} \\ j_3 & j & j_{23} \end{Bmatrix} = (-1)^{j_1+j_2+j_3+j} W(j_1 j_2 j j_3; j_{12} j_{23}). \qquad \text{(C.61)}$$

The 6-j symbols are invariant under any permutation of their columns, and under permutations of the upper and lower arguments in any pair of columns. These correspond to the following symmetry relations among the Racah coefficients:

$$\begin{aligned} W(abcd; ef) &= W(badc; ef) = W(cdab; ef) = W(acbd; ef) \\ &= (-1)^{e+f-a-d} W(ebcf; ad) = (-1)^{e+f-b-c} W(aefd; bc). \end{aligned} \qquad \text{(C.62)}$$

If one of the arguments is equal to zero, the Racah coefficients assume a very simple form. Thus,

$$W(abcd; e0) = (-1)^{a+b-e} \delta_{ac} \delta_{bd} / [(2a+1)(2b+1)]^{1/2}. \qquad \text{(C.63)}$$

The Racah coefficients and the associated with it 6-j symbols have found a wide use in a variety of problems in physics [7, 9, 13, 14]. Tables of Racah coefficients and 6-j symbols have been published, which cover a fairly wide range of arguments [15, 16].

The 6-j symbols occur in the coupling of three angular momenta. Similarly, 9-j symbols [17] make their appearance in problems which involve four angular momenta. These can be expressed as a sum of products of three 6-j symbols. In the general case, in which $n+1$ angular momenta are coupled, the transformation matrix can be expressed as a $3n-j$ symbol; a more detailed account of $3n-j$ symbols is given in Ref. [10].

C.3 Tensor Representations

C.3.1 Construction of a Tensor Representation

Consider a set of n orthonormal functions ψ_i. This set can be regarded as a vector in an n-dimensional vector space. For instance, in the central field the functions ψ_i can be the eigenfunctions ψ_{lm} of the orbital momentum L. In this case they are defined in $(2l+1)$-dimensional vector space. Under a unitary transformation of the space U a vector with components ψ_i becomes a new vector whose components are given by

$$\psi'_i = \sum_{k=1}^{n} u_{ik}\psi_k. \tag{C.64}$$

Let us form products of the components of two such vectors and apply the transformation U to both vectors. As a result, the products transform according to the following rule:

$$\psi'_{i_1}\psi'_{i_2} = \sum_{k_1 k_2} u_{i_1 k_1} u_{i_2 k_2} \psi_{k_1}\psi_{k_2}. \tag{C.65}$$

The matrix of this transformation is the direct product of the transformation matrices (C.64). The n^2 quantities $\psi_{i_1}\psi_{i_2}$ constitute a *second rank tensor* defined in the n-dimensional vector space. We denote the components of this tensor by $T_{i_1 i_2}$. In a similar way, we may form an Nth *rank tensor*, defined as the set of n^N components

$$T_{i_1 i_2 \cdots i_N} = \psi_{i_1}\psi_{i_2}\cdots\psi_{i_N}, \tag{C.66}$$

which under transformations of the vector space (C.64) transforms according to the rule

$$T'_{i_1 i_2 \ldots i_N} = \sum_{k_1, k_2, \ldots k_N} u_{i_1 k_1} u_{i_2 k_2} \ldots u_{i_N k_N} T_{k_1 k_2 \cdots k_N}. \tag{C.67}$$

A transformation of the n-dimensional space U therefore induces a transformation

$$\prod_N (U) = \underbrace{U \times U \times \cdots \times U}_{N} \tag{C.68}$$

in the n^N-dimensional space of Nth rank tensors. The matrices, which correspond to the transformation $\prod_N (U)$, form an Nth rank *tensor representation* of the unitary group $\mathbf{U}_n$. This representation is a direct product of N n-dimensional representations of $\mathbf{U}_n$ and it must be reducible, see Section A.2.7.

If one now introduces the arguments upon which the functions ψ_i depend, the tensor (C.66) can be written as a function of these arguments:

$$T_{i_1 i_2 \ldots i_N}(1, 2, \ldots, N) = \psi_{i_1}(1)\psi_{i_2}(2) \ldots \psi_{i_N}(N), \tag{C.69}$$

where the numbers 1,2,…,N enumerate the sets of arguments for each of the functions ψ_i.[10]

C.3.2 *Reduction of a Tensor Representation into Reducible Components*

It follows from the form of the transformation (C.67) that if one carries out a permutation $\bar{P}$ of the indices of the tensor (C.69), the new tensor obeys the same transformation law as the original tensor. This can be shown by taking a second rank tensor as an example:

$$\left(\bar{P}_{12} T_{i_1 i_2}\right)' = T'_{i_2 i_1} = \sum_{k_2, k_1} u_{i_2 k_2} u_{i_1 k_1} T_{k_2 k_1} = \sum_{k_1, k_2} u_{i_1 k_1} u_{i_2 k_2} \bar{P}_{12} T_{k_1 k_2}. \tag{C.70}$$

Hence, the two operations of carrying out a unitary transformation and of permuting the indices commute. If we form from the tensor components linear combinations which possess definite symmetries with respect to permutations of the indices, then under unitary transformations these linear combinations transform only among themselves. Such linear combinations can be obtained by applying the Young operators (B.33) to the arguments of the tensor (C.69):

$$\omega_{rt}^{[\lambda]} T_{i_1 i_2 \cdots i_N}(1, 2, \ldots, N). \tag{C.71}$$

The letter t in such tensors characterizes their symmetry with respect to permutations of the indices and the letter r their symmetry with respect to permutations of the arguments (see Appendix B). If (C.71) is subjected to a unitary transformation (C.68), then because of the commutation relations

$$\Pi_N(U)\omega_{rt}^{[\lambda]} = \omega_{rt}^{[\lambda]}\Pi_N(U), \tag{C.72}$$

[10] If $\psi_i(k)$ denotes the wave function for the k-th particle in a state characterized by the set of quantum numbers i, then (C.69) has the physical sense of a wavefunction for a system of N noninteracting particles.

we obtain

$$\begin{aligned}
&\Pi_N(U)\omega_{rt}^{[\lambda]}T_{i_1i_2\ldots i_N}(1,\ldots,N)\\
&=\omega_{rt}^{[\lambda]}\Pi_N(U)T_{i_1i_2\ldots i_N}(1,\ldots,N)\\
&=\sum_{k_1,\ldots,k_N}u_{i_1k_1}u_{i_2k_2}\ldots u_{i_Nk_N}\omega_{rt}^{[\lambda]}T_{k_1k_2\cdots k_N}(1,\ldots,N).
\end{aligned}\tag{C.73}$$

The components of a tensor of the form (C.71) with fixed $[\lambda]$, r, and t and with running indices i_1, i_2, …, i_N therefore transform among themselves under unitary transformations. However, not all of the components are independent. The number of independent components of a symmetrized tensor is determined by the dimension of the irreducible representation of the group $\mathbf{U}_n$, characterized by the symmetry diagram $[\lambda]$. We denote this representation by $U_n^{[\lambda]}$ and its dimension by $\delta_\lambda(n)$. A rule for finding $\delta_\lambda(n)$ can be obtained from the method of determining a basis for the representation $U_n^{[\lambda]}$.

First we note that there are only representations $U_n^{[\lambda]}$ for which the Young diagrams do not have any rows or columns longer than n. This is due to the fact that not more than n of the N indices of a tensor $T_{i_1i_2\cdots i_N}$ can be distinct. For a given Young diagram $[\lambda]$ there are f_λ^2 operators $\omega_{rt}^{[\lambda]}$. The symmetry of a tensor (C.71) under permutations of its arguments is characterized by the Young tableau r. For a given r, we obtain a set of components that transform into each other under unitary transformations. In all there will be f_λ such sets, differing from one another in the tableau r. A basis for an irreducible representation $U_n^{[\lambda]}$ can be constructed from each set of components. It should be noted that tensor components that are obtained by the action of $\omega_{r_0t}^{[\lambda]}$ with fixed r_0 upon the $T_{i_1i_2\ldots i_N}$, and that differ from one another only in a permutation of the indices, are linearly dependent. Thus, letting

$$T_{i_1i_2\ldots i_N}=\overline{P}T^0_{i_1i_2\ldots i_N},$$

then, because of relations (B.46) and (B.51),

$$\omega_{r_0t_o}^{[\lambda]}T_{p_1p_2\ldots p_N}=\omega_{r_0t_o}^{[\lambda]}\overline{P}T^0_{i_1i_2\ldots i_N}=\overline{P}\omega_{r_0t_o}^{[\lambda]}T^0_{i_1i_2\ldots i_N}=\sum_u\Gamma_{ut_o}^{[\lambda]}(P)\omega_{r_0u}^{[\lambda]}T^0_{i_1i_2\ldots i_N}.\tag{C.74}$$

The following procedure can therefore be used to find the dimension of an irreducible representation $U_n^{[\lambda]}$. We first pick out all the tensor components that do not transform into each other under permutations of the indices. We apply to them in succession the f_λ operators $\omega_{r_0t}^{[\lambda]}$ with a fixed first index. This may give a zero result in some cases. The dimension $\delta_\lambda(n)$ of the irreducible representation is then given by the number of nonzero components of the symmetrized tensor. Examples of this procedure will be given below.

The irreducible representations of the unitary group $\mathbf{U}_n$, which occur in the decomposition of an Nth rank tensor representation, are thus characterized by Young diagrams consisting of N cells, the number of cells in a column not exceeding n. The number of times that each irreducible representation of the unitary group with symmetry diagram $[\lambda]$ occurs in a decomposition is equal to the dimension of the irreducible representation of the permutation group corresponding to the same symmetry diagram $[\lambda]$.[11] As a result of the reduction of a tensor representation, the n^N basis functions can be schematically arranged in a plane diagram in the form of a series of rectangles, each rectangle being characterized by a particular Young diagram $[\lambda]$ and contains $f_\lambda \delta_\lambda(n)$ functions. Functions, which lie on the same row of a rectangle, transform into each other under unitary transformations, and functions lying in the same column transform into each other under permutations of the arguments. It is obvious that

$$\sum_{\lambda} f_\lambda \delta_\lambda(n) = n^N.$$

A very simple example is provided by the decomposition into irreducible components of a second rank tensor representation in a two-dimensional space. One can form from the four components

$$T_{\alpha\alpha},\ T_{\beta\beta},\ T_{\alpha\beta},\ T_{\beta\alpha},$$

three symmetric combinations $\omega^{[2]} T_{i_1 i_2}$:

$$T_{\alpha\alpha},\ T_{\beta\beta},\ T_{\alpha\beta} + T_{\beta\alpha},$$

and a single antisymmetric combination $\omega^{[1^2]} T_{i_1 i_2}$

$$T_{\alpha\beta} - T_{\beta\alpha}.$$

As a more complicated example, we consider a third rank tensor representation in a three-dimensional space. The 27 components can be divided into three types: (a) those with all indices identical, (b) those with two of the indices identical, and

[11] These results were first obtained by Weyl [18]. The derivation given here, see also Ref. [2], differs from the original by the use of the Young operators $\omega_{rt}^{[\lambda]}$.

(c) those with all indices different. We write these out, placing in the same column the components that transform into each other under permutation:

$$
\begin{array}{llllllllllllll}
\text{(a)} & T_{xxx} & T_{yyy} & T_{zzz} & \text{(b)} & T_{xxy} & T_{xxz} & T_{yyx} & T_{yyz} & T_{zzx} & T_{zzy} & \text{(c)} & T_{xyz} \\
 & & & & & T_{xyx} & T_{xzx} & T_{yxy} & T_{yzy} & T_{zxz} & T_{zyz} & & T_{yxz} \\
 & & & & & T_{yxx} & T_{zxx} & T_{xyy} & T_{zyy} & T_{xzz} & T_{yzz} & & T_{xzy} \\
 & & & & & & & & & & & & T_{zyx} \\
 & & & & & & & & & & & & T_{yzx} \\
 & & & & & & & & & & & & T_{zxy}
\end{array}
. \qquad (C.75)
$$

The three Young diagrams with three cells, [3], [21], and [1^3], are all allowed, and consequently there will be three nonequivalent irreducible representations in the decomposition of the tensor representation. We now determine their dimensions.

When $[\lambda]=[3]$, there is just a single Young operator $\omega^{[3]}$. The independent linear combinations can be obtained by applying $\omega^{[3]}$ to the components in the first row of (C.75). We obtain, in all, 10 such combinations, and therefore $\delta_{[3]}(3)=10$.

When $[\lambda]=[21]$, there are four Young operators $\omega_{rt}^{[21]}$; these are explicitly written in Eq. (B.42). Young operators which differ in the index r produce tensors which belong to different bases for an irreducible representation of the unitary group. The independent tensor components that belong to a single basis are found by applying operators $\omega_{r1}^{[21]}$ and $\omega_{r2}^{[21]}$ to the components in the first row of (C.75). The application of $\omega_{rt}^{[21]}$ to the components (a) gives zero. By applying $\omega_{r1}^{[21]}$ to the six components in the first row of (b) in (C.75), we obtain six linearly independent basis functions (the application of $\omega_{r2}^{[21]}$ to these components gives zero, since the components are symmetric in the first two indices). We obtain a further two basis functions by applying $\omega_{r1}^{[21]}$ and $\omega_{r2}^{[21]}$ to T_{xyz}. The dimension of the representation is therefore $\delta_{[21]}(3)=8$. In all, two such representations occur in the desired decomposition, corresponding to the two tableaux r, since $f_{[21]}=2$.

Finally, there is only one way of forming an antisymmetric tensor. This is achieved by applying the antisymmetrizing operator $\omega^{[1^3]}$ to a component in which all three indices are different, for example, to T_{xyz}. The dimension of the representation is therefore given by $\delta_{[1^3]}(3)=1$.

The 27-dimensional space of a third rank tensor can thus be decomposed into one 10-dimensional, two 8-dimensional, and a single 1-dimensional irreducible subspace. By representing each basis function of an irreducible subspace as a point on a plane diagram, we obtain three rectangles. Functions which lie on the same row of a rectangle transform into each other under unitary transformations, and functions in the same column transform into each other under permutations of their arguments.

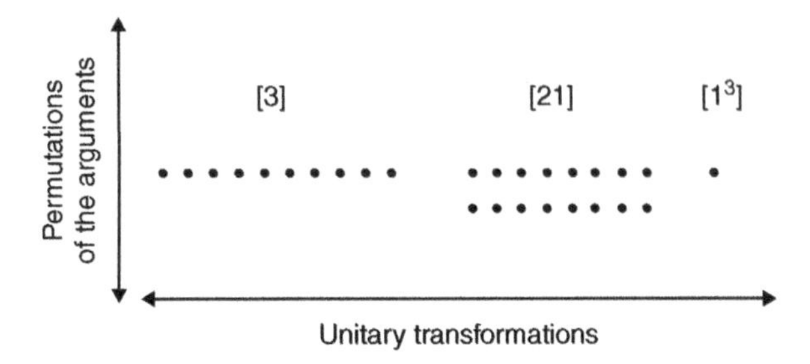

A general formula can be derived which gives the dimension of an irreducible representation of $\mathbf{U}_n$ in terms of the characteristic parameters of the associated Young diagram $[\lambda]$ [19, 20]:

$$\delta_\lambda(n) = \frac{\prod_{i<j}(h_i - h_j)}{(n-1)!(n-2)!\cdots 1!}, \quad h_i = \lambda^{(i)} + n - i. \tag{C.76}$$

In this equation i and j run from 1 to n, and $\lambda^{(i)}$ denotes the number of cells on the ith row of the Young diagram. If the number of rows in the diagram m is less than n, then $\lambda^{(i)}$ in the formula for h_i must be put equal to zero when $i > m$.

For the symmetric representation $[\lambda] = [N]$, formula (C.76) assumes the form

$$\delta_{[N]}(n) = \frac{(n+N-1)!}{[N!(n-1)!]}, \tag{C.77}$$

and for the antisymmetric representation $[\lambda] = [1^N]$, formula (C.76) becomes

$$\delta_{[1^N]}(n) = \frac{n!}{[N!(n-N)!]}. \tag{C.78}$$

Expression (C.77) coincides with the formula for the number of ways of distributing N particles among n single-particle states in the Bose–Einstein statistics, and the expression (C.78) with the analogous formula in the Fermi–Dirac statistics. This concurrence is not accidental, but arises from the fact that in the case of the Bose–Einstein statistics the total wavefunction characterizing the state of a system of N particles must be symmetric under permutations of the particles and must be antisymmetric in the case of the Fermi–Dirac statistics.

C.3.3 Littlewood's Theorem

Littlewood [21] has proved a theorem that enables one to determine which irreducible representations $\Gamma^{[\lambda]}$ of a group π_N can be formed from the direct product of two irreducible representations $\Gamma^{[\lambda_1]}$ and $\Gamma^{[\lambda_2]}$ of the groups π_{n_1} and $\pi_{n_2}(n_1+n_2=N)$, that is, on enlarging the subgroup $\pi_{n_1}\times\pi_{n_2}$ to the group π_N. In contrast to the formation of the more usual direct product, this process is known as forming the *outer product*, and is denoted by the symbol ⊗. The decomposition of an outer product into irreducible components can be easily carried out with the aid of Littlewood's theorem:

$$\Gamma^{[\lambda_1]}\otimes\Gamma^{[\lambda_2]}\doteq\sum_{\lambda}a(\lambda_1,\lambda_2,\lambda)\Gamma^{[\lambda]}. \tag{C.79}$$

Since an *N*th rank tensor which has been symmetrized with respect to a Young diagram $[\lambda]$ can simultaneously serve as a basis for both the permutation and the unitary groups (see the previous section), Littlewood's theorem can also be used to find which irreducible representations $U_n^{[\lambda]}$ occur in the decomposition of the direct product

$$U_n^{[\lambda_1]}\times U_n^{[\lambda_2]}\doteq\sum_{\lambda}a(\lambda_1,\lambda_2,\lambda)U_n^{[\lambda]}. \tag{C.80}$$

The summation on the right of this equation includes only those Young diagrams of (C.79) in which the length of the columns does not exceed n. This restriction automatically ceases to arise when $n\geq N$.

Before stating Littlewood's theorem, we introduce the concept of a *lattice permutation*. Consider an expression of the form $x_1^{n_1}x_2^{n_2}x_3^{n_3}\cdots(n_1\geq n_2\geq n_3\geq\cdots)$. A reordering of the elements x_1, x_2, x_3, …, is called a lattice permutation if among any first m terms of such a permutation, the numbers of times x_1 occurs is equal to or greater than the number of times x_2 occurs, which is equal to or greater than the number of times x_3 occurs, and so on.

As an example, the six possible lattice permutations of the expression $x_1^3x_2x_3$ are

$$\begin{array}{lll} x_1^3x_2x_3, & x_1^2x_2x_1x_3, & x_1^2x_2x_3x_1, \\ x_1x_2x_1^2x_3 & x_1x_2x_1x_3x_1, & x_1x_2x_3x_1^2 \end{array}$$

We now proceed to state the theorem:

The possible Young diagrams $[\lambda]$, *which can be constructed from the diagrams* $[\lambda_1]\equiv\left[\lambda_1^{(1)}\lambda_1^{(2)}\ldots\right]$ *and* $[\lambda_2]\equiv\left[\lambda_2^{(1)}\lambda_2^{(2)}\ldots\right]$, *are found by consecutively adding to the diagram* $[\lambda_1]$ *in all possible ways* $\lambda_2^{(1)}$ *cells containing the index* α, $\lambda_2^{(2)}$ *cells containing the index β, etc., such that in the augmented diagram the indices which have been added satisfy two conditions: (a) two identical indices may not occur in the same column, and (b) when for each obtained [λ] all the indices, which have been added, is read from right to left in consecutive rows, we obtain a lattice permutation of the expression* $\alpha^{\lambda_2^{(1)}}, \beta^{\lambda_2^{(2)}}, \ldots$.

EXAMPLE 1. $[1^3]\otimes[1^2]$:

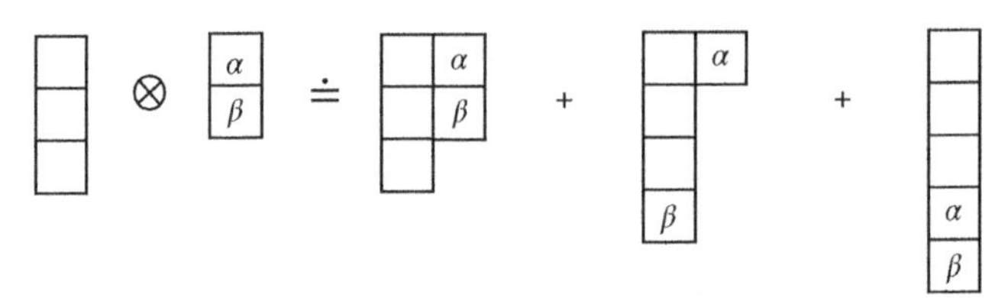

That is, $[1^3]\otimes[1^2]\doteq[2^21]+[21^3]+[1^5]$.

EXAMPLE 2. $[21]\otimes[21]$:

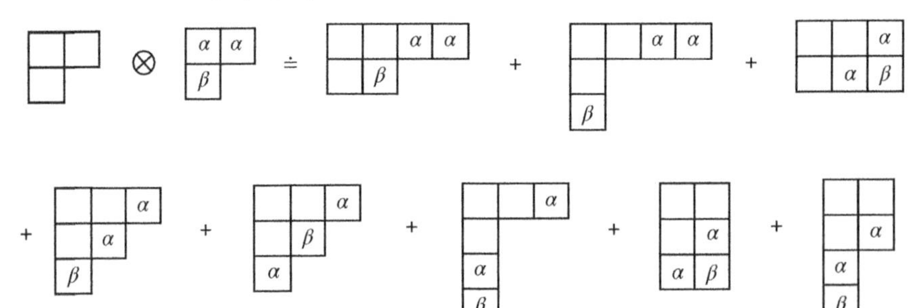

On collecting similar terms together, we obtain

$$[21]\otimes[21]\doteq[42]+[41^2]+[3^2]+2[321]+[31^3]+[2^3]+[2^21^2].$$

The decomposition (C.80) must coincide with this for any $n \geq N$. This circumstance can be used to check a decomposition carried out by using Littlewood's theorem, namely, the condition

$$\delta_{\lambda_1}(n)\delta_{\lambda_2}(n)=\sum_{\lambda} a(\lambda_1,\lambda_2,\lambda)\delta_\lambda(n), \qquad \text{(C.81)}$$

must be fulfilled. The dimensions of the representations $\delta_\lambda(n)$ are calculated by formula (C.76). We can verify that Eq. (C.81) is satisfied for Example 1. From formula (C.76) we find

$$
\begin{array}{llllll}
[\lambda]: & [1^3] & [1^2] & [2^2 1] & [21^3] & [1^5] \\
\delta_\lambda(5): & 10 & 10 & 75 & 24 & 1
\end{array}
$$

which gives

$$10 \cdot 10 = 75 + 24 + 1,$$

in complete agreement with (C.81). Eq. (C.81) is also satisfied for large n. Indeed for $n = 7$ we obtain

$$35 \cdot 21 = 490 + 224 + 21.$$

Littlewood's theorem can also be used to find the representations $\Gamma^{[\lambda_1]} \times \Gamma^{[\lambda_2]}$ into which $\Gamma^{[\lambda]}$ decomposes on reduction to the subgroup $\boldsymbol{\pi}_{n_1} \times \boldsymbol{\pi}_{n_2}$. For this purpose, it is necessary to apply the theorem to every possible term in the decomposition[12]

$$\Gamma^{[\lambda]} \doteq \sum_{\lambda_1, \lambda_2} a(\lambda, \lambda_1, \lambda_2) \Gamma^{[\lambda_1]} \times \Gamma^{[\lambda_2]}, \tag{C.82}$$

and to reject those terms that do not yield the given $[\lambda]$. This method is quicker than a decomposition carried out with the aid of character tables for $\boldsymbol{\pi}_N$ and $\boldsymbol{\pi}_{n_1} \times \boldsymbol{\pi}_{n_2}$ if n_2 is not too large.

For example, let us find the representations that arise on reducing $\Gamma^{[\lambda]} = \Gamma^{[32]}$ with respect to the subgroup $\boldsymbol{\pi}_3 \times \boldsymbol{\pi}_2$. The diagram $[\lambda_1]$ is restricted to being [3] or [21], since $[1^3]$ is not contained in [32]. The desired decomposition is

$$[32] \doteq [3] \times [2] + [21] \times [2] + [21] \times [1^2].$$

The representation $\Gamma^{[3]} \times \Gamma^{[1^2]}$ does not appear, since by Littlewood's theorems, one cannot construct the diagram [32] from the diagrams [3] and $[1^2]$.

C.3.4 The Reduction of $\mathrm{U}_{2j+1} \rightarrow \mathrm{R}_3$

Let us consider the $2j + 1$ functions $\psi_m^{(j)}$, which form a basis for an irreducible representation $D^{(j)}$ of the rotation group in three dimensions $\mathbf{R}_3$. The basis functions transform into each other under rotations:

---

[12] By Frobenius' theorem (see Chapter 4, Section 4.2.1) it follows that the coefficients $a(\lambda, \lambda_1, \lambda_2)$ in Eq. (C.82) are equal to the coefficients $a(\lambda_1, \lambda_2, \lambda)$ in Eq. (C.79).

$$R\psi_m^{(j)} = \sum_{m'} D_{m'm}^{(j)}(R)\psi_{m'}^{(j)}. \tag{C.83}$$

The set of $2j+1$ functions $\psi_m^{(j)}$ can be regarded as a vector in a $(2j+1)$-dimensional space. Every rotations in three-dimensional space thus generates some unitary transformation in the $(2j+1)$-dimensional space which is implemented by the matrices of the representation $D^{(j)}$. The group of such transformations forms a subgroup of the group of all unitary transformations $\mathbf{U}_n$.

Under rotations of the three-dimensional space the product of functions

$$\psi_{m_1}^{(j)}\psi_{m_2}^{(j)}\ldots\psi_{m_N}^{(j)}, \tag{C.84}$$

transforms according to the $m_1 m_2 \ldots m_N$ column of the direct product

$$\underbrace{D^{(j)} \times D^{(j)} \times \cdots \times D^{(j)}}_{N}.$$

The irreducible representations, which occur in the decomposition of this direct product, are found by successive applications of the decomposition (C.43). On the other hand, the product (C.84) transforms under unitary transformations as an Nth rank tensor representation of the group $\mathbf{U}_{2j+1}$. The irreducible representations which occur in its decomposition $U_{2j+1}^{[\lambda]}$ are characterized by Young diagrams with N cells, the columns of which do not exceed $2j+1$ cells in length. Since the group $\mathbf{R}_3$ is a subgroup of $\mathbf{U}_{2j+1}$, the irreducible representations $U_{2j+1}^{[\lambda]}$ in general become reducible upon restricting the operations to those of $\mathbf{R}_3$ and hence the $U_{2j+1}^{[\lambda]}$ split into irreducible representations $D^{(j)}$. In classifying the states of a system of N identical particles, it is often important to know which $D^{(j)}$ occur in the decomposition

$$U_{2j+1}^{[\lambda]} \doteq \sum_J a^{(J)} D^{(J)}, \tag{C.85}$$

and in order to discover this, we make use of a recursive method proposed by Jahn [22].

We note first of all that the one-dimensional representation $U_{2j+1}^{[1^{2j+1}]}$ must correspond to the representations $D^{(0)}$. This is because all the indices m_i in the nonzero components of the tensor $T_{m_1 \ldots m_{2j+1}}^{[1^{2j+1}]}$ must be different, and hence there is in fact only one nonzero component. The tensor therefore corresponds to an angular momentum $J=0$. It can be shown that as far as the decomposition (C.85) is concerned, the representations $U_{2j+1}^{[\lambda]}$ with the Young diagrams

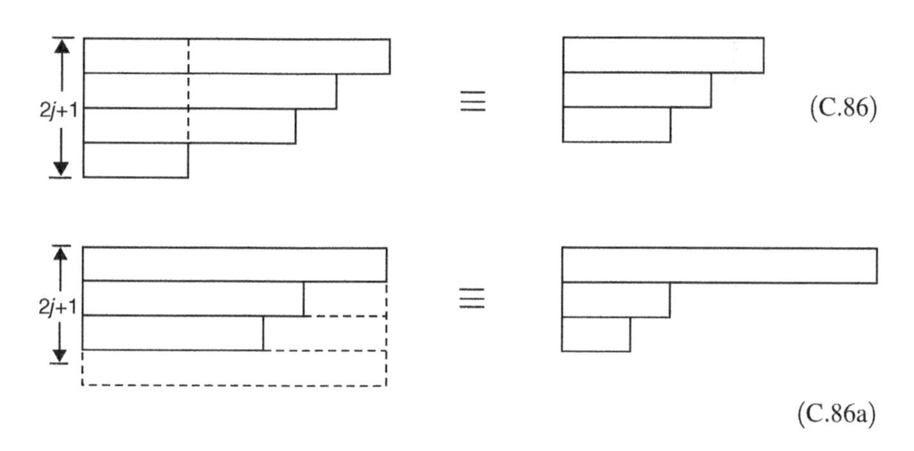

(C.86)

(C.86a)

are equivalent. Relation (C.86a) depends upon the fact that a Young diagram whose columns are all of length $2j+1$ corresponds to an angular momentum $J=0$. Hence, the two Young diagrams which make up the rectangle in (C.86a) must correspond to the same values of J otherwise one would not obtain $J=0$ on coupling the two corresponding vectors.

We consider the method of determining the form of the decomposition (C.85) for the group $\mathbf{U}_3$, that is, for $j=1$. The representation $U_3^{[1]}$ is generated by the three functions $\psi_1^{(1)}$, $\psi_0^{(1)}$, and $\psi_{-1}^{(1)}$ which under rotations transform according to the irreducible representation $D^{(1)}$. Hence, we have

$$U_3^{[1]} \doteq D^{(1)}. \tag{C.87}$$

When $N=2$, the two irreducible representations $U_3^{[2]}$ and $U_3^{[1^2]}$ occur in the decomposition of the tensor representation. However, when the direct product $D^{(1)} \times D^{(1)}$ is reduced, we obtain three irreducible representations $D^{(2)}$, $D^{(1)}$, and $D^{(0)}$. In order to find the correspondence between these representations, we make use of the symmetry of a basis function of a representation $D^{(J)}$, $\psi_M^{(J)}(j(1)j(2))$, with respect to a permutation of the order in which the angular momenta $j(i)$ are coupled. According to Eq. (C.49), the function is multiplied by $(-1)^{2j-J}=(-1)^J$ (for j integral) under such a permutation; that is, the function is symmetric for J even and antisymmetric for J odd. Since the basis functions for $U_3^{[2]}$ are symmetric with respect to a permutation of the indices, whereas those for $U_3^{[1^2]}$ are antisymmetric, we obtain

$$U_3^{[2]} \doteq D^{(2)} + D^{(0)}, \quad U_3^{[1^2]} \doteq D^{(1)}. \tag{C.88}$$

Proceeding further with this decomposition, we form from the Young diagram [2] all the possible diagrams with three cells. We find the representations $D^{(J)}$ that correspond to these diagrams by means of the triangle rule (C.44). The diagram $[1^2]$ is dealt with similarly:

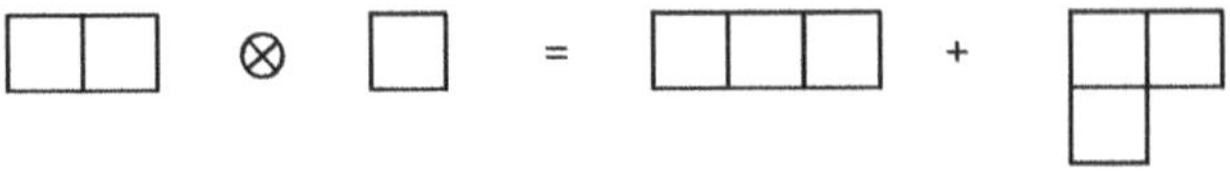

$$\left(D^{(2)}+D^{(0)}\right)\times D^{(1)} \doteq D^{(3)}+D^{(2)}+2D^{(1)}. \tag{C.89}$$

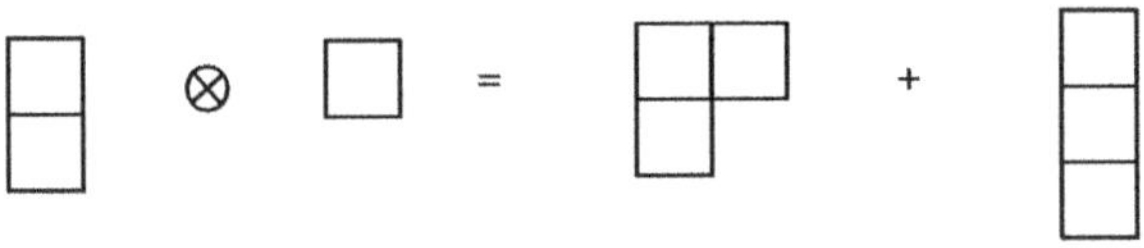

$$D^{(1)}\times D^{(1)} \doteq D^{(2)}+D^{(1)}+D^{(0)}.$$

As a result we obtain the two equations

$$\begin{aligned} U_3^{[3]}+U_3^{[21]} &\doteq D^{(3)}+D^{(2)}+2D^{(1)}, \\ U_3^{[21]}+U_3^{[1^3]} &\doteq D^{(2)}+D^{(1)}+D^{(0)}. \end{aligned} \tag{C.90}$$

Since only $D^{(0)}$ may correspond to $U^{[1^3]}$, we arrive, with the aid of Eq. (C.90) at the desired decompositions:

$$U_3^{[3]} \doteq D^{(3)}+D^{(1)}, \quad U_3^{[21]} \doteq D^{(2)}+D^{(1)}, \quad U_3^{[1^3]} \doteq D^{(0)}. \tag{C.91}$$

For $N=4$, similarly to (C.89) and using Eq. (C.91), we have

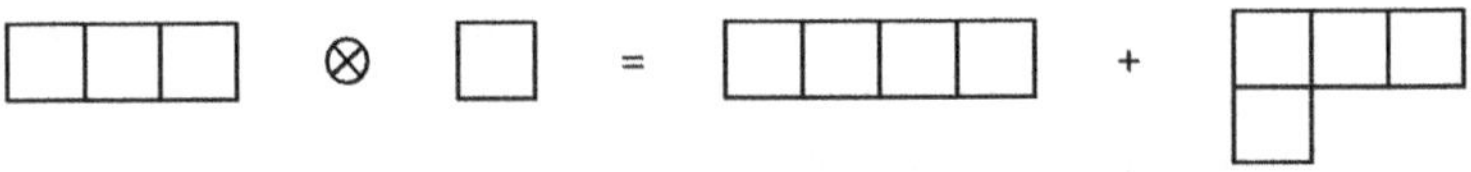

$$\left(D^{(3)}+D^{(1)}\right)\times D^{(1)} \doteq D^{(4)}+D^{(3)}+2D^{(2)}+D^{(1)}+D^{(0)}$$

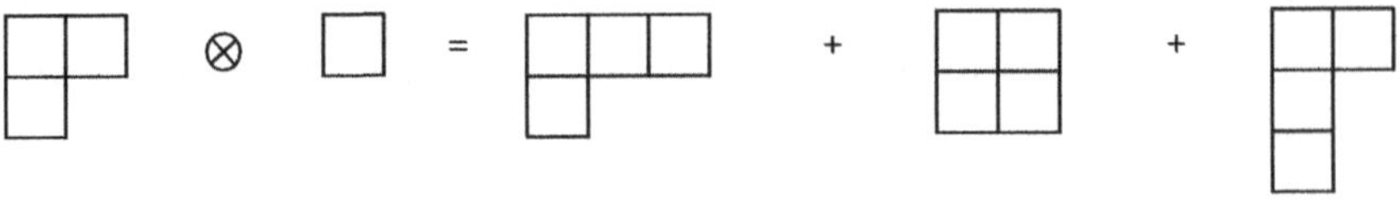

(C.92)

$$\left(D^{(2)}+D^{(1)}\right)\times D^{(1)} \doteq D^{(3)}+2D^{(2)}+2D^{(1)}+D^{(0)},$$

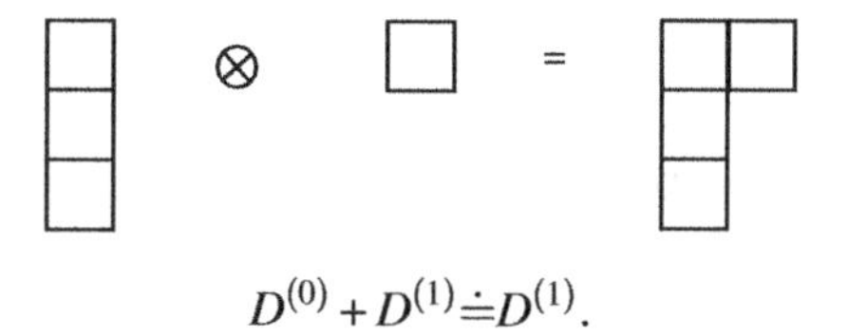

$$D^{(0)} + D^{(1)} \doteq D^{(1)}.$$

According to (C.86a), the representation $U_3^{[2^2]}$ possesses the same J structure as $U_3^{[2]}$, and $U_3^{[21^2]}$ the same J structure as $U_3^{[1^2]}$. Taking this into account, we obtain from (C.92)

$$\begin{aligned} &U_3^{[4]} \doteq D^{(4)} + D^{(2)} + D^{(0)}, \quad U_3^{[31]} \doteq D^{(3)} + D^{(2)} + D^{(1)}, \\ &U_3^{[2^2]} \doteq D^{(2)} + D^{(0)}, \qquad\quad U_3^{[21^2]} \doteq D^{(1)}. \end{aligned} \tag{C.93}$$

The J structure of any representation $U_{2j+1}^{[\lambda]}$ can be found similarly. It is only necessary to bear in mind that if j is half-integral, an odd value of J corresponds to the representation $U_{2j+1}^{[2]}$ and an even value to $U_{2j+1}^{[1^2]}$. In next section, we represent the results of reducing the representations $U_{2j+1}^{[\lambda]}$ for $N=1$ to $N=4$ and $j=1$ to $j=3$.

The irreducible representations of the group $\mathbf{U}_{2j+1}$ remain irreducible on passing to the subgroup $\mathbf{SU}_{2j+1}$ [20]. However, some of the irreducible representations may as a result become equivalent. It turns out that the conditions for this to occur are the same as the conditions (C.86). The reduction of the irreducible representations $\mathbf{SU}_{2j+1}$ to those of $\mathbf{R}_3$ of course proceeds similarly to that of $\mathbf{U}_{2j+1} \to \mathbf{R}_3$. However, in the case of $\mathbf{SU}_{2j+1}$ there is now a unique correspondence between the various irreducible representations and their J structure.

The reduction of the unitary group in two-dimensional space $\mathbf{U}_2 \to \mathbf{R}_3$, does not cause any splitting of the representations $U_2^{[\lambda]}$. This is connected with the fact that the groups $\mathbf{SU}_2$ and $\mathbf{R}_3$ are isomorphic, and hence there is a one-to-one correspondence between the irreducible representations of these two groups.[13] This can easily be verified by carrying out the procedure of successively adding cells to the Young diagrams. The diagram for the representations $U_2^{[\lambda]}$ may not have more than two cells in a column, that is, they consist of two rows. The angular momentum J that corresponds to a diagram $[\lambda]$ is determined by the lengths of the rows $\lambda^{(1)}$ and $\lambda^{(2)}$ as follows:

$$J = \frac{1}{2}\left(\lambda^{(1)} - \lambda^{(2)}\right), \tag{C.94}$$

[13] This circumstance is of great importance in the so-called quantum chemistry without spin and in the classification of the Pauli-allowed states in many-electron system, see Chapters 2 and 4.

thus

$$U_2^{[41]} \leftrightarrow D^{(\frac{3}{2})}, \quad U_2^{[31]} \leftrightarrow D^{(1)}, \quad U_2^{[2^2]} \leftrightarrow D^{(0)}.$$

C.4 Tables of the Reduction of the Representations $U_{2j+1}^{[\lambda]}$ to the Group $\mathbf{R}_3$

The tables of the reduction of the representations $U_{2j+1}^{[\lambda]}$ for states of systems with one to four particles and $j=1$ to $j=3$ were calculated in book [2] and are represented (Tables C.1, C.2, C.3, C.4, C.5, and C.6). For larger numbers of particles, the reduction can be carried out by the method described in the previous section.

Table C.1 $j=\frac{1}{2}$

N	$[\lambda]$	J	$\delta_\lambda(2)$
1	[1]	1/2	2
2	[2]	1	3
	$[1^2]\equiv[0]$	0	1
3	[3]	3/2	4
	$[21]\equiv[1]$	1/2	2
4	[4]	2	5
	$[31]\equiv[2]$	1	3
	$[2^2]\equiv[0]$	0	1

Table C.2 $j=1$

N	$[\lambda]$	J	$\delta_\lambda(3)$
1	[1]	1	3
2	[2]	0, 2	6
	$[1^2]$	1	3
3	[3]	1, 3	10
	[21]	1, 2	8
	$[1^3]\equiv[0]$	0	1
4	[4]	0, 2, 4	15
	[31]	1, 2, 3	15
	$[2^2]\equiv[2]$	0, 2	6
	$[21^2]\equiv[1^2]$	1	3

Table C.3 $j = \frac{3}{2}$

N	$[\lambda]$	J	$\delta_\lambda(4)$
1	[1]	3/2	4
2	[2]	1, 3	10
	$[1^2]$	0, 2	6
3	[3]	3/2, 5/2, 9/2	20
	[21]	1/2, 3/2, 5/2, 7/2	20
	$[1^3] \equiv [1]$	3/2	4
4	[4]	0, 2, 3, 4, 6	35
	[31]	$1^2, 2, 3^2, 4, 5$	45
	$[2^2]$	$0, 2^2, 4$	20
	$[21^2]$	1, 2, 3	15
	$[1^4] \equiv [0]$	0	1

Table C.4 $j = 2$

N	$[\lambda]$	J	$\delta_\lambda(5)$
1	[1]	2	5
2	[2]	0, 2, 4	15
	$[1^2]$	1, 3	10
3	[3]	0, 2, 3, 4, 6	35
	[21]	$1, 2^2, 3, 4, 5$	40
	$[1^3] \equiv [1^2]$	1, 3	10
4	[4]	$0, 2^2, 4^2, 5, 6, 8$	70
	[31]	$1^2, 2^2, 3^3, 4^2, 5^2, 6, 7$	105
	$[2^2]$	$0^2, 2^2, 3, 4^2, 6$	50
	$[21^2]$	$1^2, 2, 3^2, 4, 5$	45
	$[1^4] \equiv [1]$	2	5

Table C.5 $j = \frac{5}{2}$

N	$[\lambda]$	J	$\delta_\lambda(6)$
1	[1]	5/2	6
2	[2]	5, 3, 1	21
	$[1^2]$	4, 2, 0	15
3	[3]	3/2, 5/2, 7/2, 9/2, 11/2, 15/2	56
	[21]	$1/2, 3/2, (5/2)^2, (7/2)^2, 9/2, 11/2, 13/2$	70
	$[1^3]$	3/2, 5/2, 9/2	20
4	[4]	$0, 2^2, 3, 4^2, 5, 6^2, 7, 8, 10$	126
	[31]	$1^3, 2^2, 3^4, 4^3, 5^4, 6^2, 7^2, 8, 9$	210
	$[2^2]$	$0^2, 2^3, 3, 4^3, 5, 6^2, 8$	105
	$[21^2]$	$1^2, 2^2, 3^3, 4^2, 5^2, 6, 7$	105
	$[1^4] \equiv [1^2]$	4, 2, 0	15

Table C.6 $j=3$

N	$[\lambda]$	J	$\delta_\lambda(7)$
1	$[1]$	3	7
2	$[2]$	0, 2, 4, 6	28
	$[1^2]$	1, 3, 5	21
3	$[3]$	$1, 3^2, 4, 5, 6, 7, 9$	84
	$[21]$	$1, 2^2, 3^2, 4^2, 5^2, 6, 7, 8$	112
	$[1^3]$	0, 2, 3, 4, 6	35
4	$[4]$	$0^2, 2^2, 3, 4^3, 5, 6^3, 7, 8^2, 9, 10, 12$	210
	$[31]$	$1^3, 2^3, 3^5, 4^4, 5^5, 6^4, 7^4, 8^2, 9^2, 10, 11$	378
	$[2^2]$	$0^2, 2^4, 3, 4^4, 5^2, 6^3, 7, 8^2, 10$	196
	$[21^2]$	$1^3, 2^2, 3^4, 4^3, 5^4, 6^2, 7^2, 8, 9$	210
	$[1^4] \equiv [1^3]$	0, 2, 3, 4, 6	35

References

[1] L.S. Pontriagin, *Continuous Groups*, Princeton University Press, Princeton, New Jersey, 1946.

[2] I.G. Kaplan, *Symmetry of Many-Electron Systems*, Academic Press, New York, 1975.

[3] G. Racah, *Ergeb. Exact. Naturwiss.* **37**, 28 (1965).

[4] L.D. Landau and E.M. Lifshitz, *Quantum Mechanics (Nonrelativistic Theory)*, 3rd edn., Pergamon Press, Oxford, 1977.

[5] I.M. Helfand, R.A. Minlos, and Z. Ya.Shapiro, *Representation of the Rotation and Lorentz Groups*, Macmillan, New York, 1963.

[6] E. Wigner, *Group Theory*, Academic Press, New York, 1959.

[7] M.E. Rose, *Elementary Theory of Angular Momentum*, John Wiley & Sons, Inc., New York, 1957.

[8] H. Bethe, *Ann. Phys. (Leipzig)* **3**, 133 (1939).

[9] A.R. Edmonds, *Angular Momentum in Quantum Mechanics*, Princeton University Press, Princeton, New Jersey, 1957.

[10] A.P. Yutsis, I.B. Levinson, and V.V. Vanagas, The Theory of Angular Momentum, in *Handbook of Molecular Physics and Quantum Chemistry*, ed. S. Wilson, John Wiley & Sons, Ltd, Chichester, 2003, vol. **1**, 390–428.

[11] G. Racah, *Phys. Rev.* **62**, 438 (1942).

[12] G. Racah, *Phys. Rev.* **63**, 367 (1943).

[13] U. Fano and G. Racah, *Irreducible Tensorial Sets*, Academic Press, New York, 1959.

[14] B.R. Judd, *Operator Techniques in Atomic Spectroscopy*, McGraw-Hill, New York, 1963.

[15] M. Rotenberg, N. Metropolis, R. Bivins, and J.K. Wooten, Jr., *The 3-j and 6-j Symbols*, MIT Press, Cambridge, Massachusetts, 1959.

[16] T. Ishidzu, H. Horie, S. Obi, M. Sata, Y. Tanabi, and S. Yanagawa, *Tables of Racah Coefficients*, Pan-Pacific Press, Tokyo, 1960.

[17] E.P. Wigner, On the matrices which reduced the Kronecker products of simply reducible groups, in *Quantum Theory of Angular Momentum*, eds. L.S. Biedenharn and H. van Dam, Academic Press, New York, 1965.

[18] H. Weyl, *The Classical Groups*, Princeton University Press, Princeton, New Jersey, 1966.

[19] F.D. Murnagan, *The Theory of Group Representations*, John Hopkins Press, Baltimore, Maryland, 1938.

[20] M. Hamermesh, *Group Theory*, Addison Wesley, Reading, Massachusetts, 1964.

[21] D.E. Littlewood, *The Theory of Group Characters and Matrix Representations of Groups*, Oxford University Press (Clarendon), London, 1940.

[22] H.A. Jahn, *Proc. Roy. Soc. Lond. A* **201**, 516 (1950).

Appendix D

Irreducible Tensor Operators

D.1 Definition

As it was shown in Section C.3.2, Appendix C, the n^N components of an Nth rank tensor transform under a unitary transformation of the n-dimensional space as a reducible representation which decomposes into irreducible components when the tensor is symmetrized according to a Young diagram. The tensors, which are symmetrized in this way, are irreducible with respect to the operations of the unitary group $\mathbf{U}_n$, that is, the components of a symmetrized tensor transform under unitary transformations only among themselves. In general, however, these tensors are not irreducible with respect to the operations of the three-dimensional rotation group $\mathbf{R}_3$. For example, according to Eq. (C.88) in Appendix C, one can form from the six components of a symmetrized second-rank tensor, which transforms according to the irreducible representation $U_3^{[2]}$, one scalar (the irreducible representation $D^{(0)}$) and five linear combinations which transform according to the irreducible representation $D^{(2)}$.

In general, a set of f_α quantities $T_i^{(\alpha)}$ is called an *irreducible tensor* of a group of linear transformations, if under the operations of the group the $T_i^{(\alpha)}$ transform according to an irreducible representation $\Gamma^{(\alpha)}$ of this group:

$$RT_i^{(\alpha)} = \sum_{k=1}^{f_\alpha} \Gamma_{ki}^{[\alpha]}(R) T_k^{(\alpha)}. \tag{D.1}$$

The Pauli Exclusion Principle: Origin, Verifications, and Applications, First Edition. Ilya G. Kaplan.

From this definition it follows that any set of basis functions for an irreducible representation can be regarded as an irreducible tensor. Thus the set of $2J+1$ spherical harmonic functions[1] $Y_{JM}(\theta, \phi)$ form an example of an irreducible tensor which belongs to the representations $D^{(J)}$ of the group $\mathbf{R}_3$. The Cartesian components of an arbitrary vector **A** form a first-rank tensor. However, their transformational properties under three-dimensional rotations are more complicated than those of the spherical components

$$A_0 = A_z, \quad A_{\pm 1} = \mp \frac{1}{\sqrt{2}}\left(\mathrm{A_x} \pm \mathrm{iA_y}\right),$$

which constitute an irreducible tensor of the representations $D^{(1)}$. When calculating matrix elements of vector quantities with basis functions of the group $\mathbf{R}_3$ it is therefore more convenient to use the spherical components of the vectors.

Three types of operation can be defined for ordinary (i.e., Cartesian) tensors $T^{(N)}$ (N denotes the rank of a tensor):

1. Addition of tensors of identical rank,

$$V_{ik}^{(2)} = T_{ik}^{(2)} + U_{ik}^{(2)}. \tag{D.2}$$

2. Tensor multiplication,

$$V_{iklmn}^{(5)} = T_{ik}^{(2)} V_{lmn}^{(3)}, \tag{D.3}$$

   which leads to a tensor whose rank is equal to the sum of the ranks of the tensors forming the product.

3. Contraction over a pair of indices, which leads to a lowering of the rank of the tensor,

$$T_i^{(1)} = \sum_k T_{ikk}^{(3)}. \tag{D.4}$$

For tensors of even rank, $N/2$ successive contractions lead to a scalar. For instance,

$$T^{(0)} = \sum_{i,k} T_{iikk}^{(4)}. \tag{D.5}$$

[1] Irreducible tensors of the group $\mathbf{R}_3$ are called *spherical tensors*.

The scalar product of two tensors is defined as a contraction over all indices:

$$V^{(0)} = \sum_{i,k} T_{ik}^{(2)*} V_{ik}^{(2)}. \tag{D.5a}$$

The addition of irreducible tensors is similar to the addition of Cartesian tensors,

$$V_i^{(\alpha)} = T_i^{(\alpha)} + U_i^{(\alpha)}. \tag{D.6}$$

However, instead of tensors multiplication or contraction, the following operation is defined for irreducible tensors:

$$V_t^{(\tau)} = \sum_{i,k} T_i^{(\alpha)} U_k^{(\beta)} \langle \alpha i, \beta k | \tau t \rangle. \tag{D.7}$$

As a result one obtains a tensor that transforms according to an irreducible representation $\Gamma^{(\tau)}$ which occurs in the decomposition of the direct product $\Gamma^{(\alpha)} \times \Gamma^{(\beta)}$. The coefficients in Eq. (D.7) are the Clebsch–Gordan coefficients, see Eq. (A.60) in Appendix A.

A tensor, which belongs to the unit (totally symmetric) representation of a group, behaves as a scalar with respect to the operations of the particular group. According to (A.57), the necessary and sufficient condition for the unit representation to occur in the decomposition of the direct product of two representations is that the representations are complex conjugates of each other. Consequently, one can never form a scalar from a product of two irreducible tensors that belong to two different representations. In the case of the group $\mathbf{R}_3$ a scalar is characterized by a value of zero for the angular momentum, $J = 0$, and is constructed from two spherical tensors with the same J. Substituting the values of the Clebsch–Gordan coefficients (C.48) in Appendix C into (D.7), we obtain the following scalar from two spherical tensors:

$$V^{(0)} = \left[\frac{(-1)^J}{(2J+1)^{1/2}} \right] \sum_m (-1)^m T_m^{(J)} U_{-m}^{(J)}. \tag{D.8}$$

Expression (D.8) without the multiplying factor in front of the sum is usually called the scalar product of two spherical tensors and is denoted by

$$\left(T^{(J)} \cdot U^{(J)} \right) = \sum_m (-1)^m T_m^{(J)} U_{-m}^{(J)}. \tag{D.9}$$

When $J = 1$ this expression coincides with the usual scalar product of two vectors expressed in spherical coordinates.

In calculating the matrix elements of any operator it is important to know according to which irreducible representations the factors in the integrand transform. This knowledge enables one, for example, immediately to distinguish nonzero matrix elements and to obtain the group theoretical selection rules. It is therefore convenient to write the operator in the matrix element in the form of a sum of operators each of which transforms according to definite irreducible representations of the group.

An irreducible tensor operator is defined (by analogy with an irreducible tensor) as a set of f_α quantities $T_i^{(\alpha)}$ whose transformation law[2] is

$$R^{-1}T_i^{(\alpha)}R = \sum_k \Gamma_{ki}^{(\alpha)}(R)T_k^{(\alpha)}. \tag{D.10}$$

The use of irreducible tensor operators facilitates the calculation of matrix elements considerably, since one can employ a whole series of useful relations [1]. These relations rest upon the Wigner–Eckart theorem.

D.2 The Wigner–Eckart Theorem

Let us consider a matrix element of an irreducible tensor operator $T_t^{(\tau)}$,

$$\langle \alpha i | T_t^{(\tau)} | \beta k \rangle \tag{D.11}$$

defined in terms of functions which transform according to the irreducible representations $\Gamma^{(\alpha)}$ and $\Gamma^{(\beta)}$ of the same group as the representation $\Gamma^{(\tau)}$. We examine the transformational properties of the function, which results by allowing $T_t^{(\tau)}$ to operate upon the function $\psi_k^{(\beta)}$ with respect to the operations of the group:

$$R\left(T_t^{(\tau)}\psi_k^{(\beta)}\right) = \left(R^{-1}T_t^{(\tau)}R\right)\left(R\psi_k^{(\beta)}\right) = \sum_{t',k'} \Gamma_{t't}^{(\tau)}(R)\Gamma_{k'k}^{(\beta)}(R)\left(T_{t'}^{(\tau)}\psi_{k'}^{(\beta)}\right).$$

The function $T_t^{(\tau)}\psi_k^{(\beta)}$ consequently transforms according to the direct product $\Gamma^{(\tau)} \times \Gamma^{(\beta)}$, and by using Clebsch–Gordan coefficients, it can be written in the form of a decomposition into basis functions for irreducible representations of the group, see Eq. (A.63) in Appendix A.

[2] Irreducible tensor operators transform under the operations of a group in the same way as irreducible tensors. The difference between Eqs. (D.10) and (D.1) is due to the fact that operators in the new and old bases are related by Eq. (A.22).

$$T_t^{(\tau)}\psi_k^{(\beta)} = \sum_{a,\mu,m} \Phi_m^{(a\mu)}(\tau\beta)\langle a\mu m|\tau t,\beta k\rangle. \quad \text{(D.12)}$$

In this equation the index a distinguishes irreducible representations that are repeated, and the symbols τ and β in the argument of the function Φ indicate that the form of Φ depends upon the basis functions in the direct product. Substituting (D.12) into (D.11), we obtain

$$\langle \alpha i|T_t^{(\tau)}|\beta k\rangle = \sum_{a,\mu,m} \langle \alpha\mu m|\tau t,\beta k\rangle \int \psi_i^{(\alpha)*}\Phi_m^{(a\mu)}(\tau\beta)dV. \quad \text{(D.13)}$$

According to the orthogonality conditions (A.78) for basis functions of irreducible representations,

$$\int \psi_i^{(\alpha)*}\Phi_m^{(a\mu)}(\tau\beta)dV = \delta_{\alpha\mu}\delta_{im}A_a(\alpha,\tau,\beta), \quad \text{(D.14)}$$

where the quantity $A_a(\alpha, \tau, \beta)$ is determined by the form of the functions $\psi_i^{(\alpha)}$ and $\Phi_m^{(a\alpha)}(\tau\beta)$, but does not depend upon i and m. On substituting (D.14) into (D.13), we obtain an analytical expression of the Wigner–Eckart theorem[3]:

$$\langle \alpha i|T_t^{(\tau)}|\beta k\rangle = \sum_{a} \langle a\alpha i|\tau t,\beta k\rangle A_a(\alpha,\tau,\beta). \quad \text{(D.15)}$$

It follows from this equation that the Clebsch–Gordan coefficients completely determine the dependence on the matrix element upon the column numbers of the irreducible representations according to which the factors in the integrand transform. The presence of these coefficients allows one immediately to obtain the conditions under which a matrix element reduces to zero. These are as follows:

The matrix element (D.15) is zero whenever the decomposition of the direct product $\Gamma^{(\tau)} \times \Gamma^{(\beta)}$ into irreducible components does not contain the representation $\Gamma^{(\alpha)}$.

Systems with differing physical structures may possess identical symmetries. The matrix elements, which occur in quantum mechanical calculations on such systems, differ only in the factor A_a, the Clebsch–Gordan coefficients being identical. Consequently, the Wigner–Eckart theorem enables one to "separate off" the symmetry properties of a system, which is being studied, from its physical structure.

[3] This theorem was proved by Wigner [2] and by Eckart [3] for the three-dimensional rotational group. By writing it in the form (D.15), Koster [4] subsequently extended the theorem to include an arbitrary finite group.

The operators which are employed in quantum mechanics are symmetric with respect to all identical particles, that is, they transform according to the unit representation $\Gamma^{[N]}$ of the permutation group. Since $\Gamma^{[N]} \times \Gamma^{[\lambda]} = \Gamma^{[\lambda]}$, the sum over a in (D.15) reduces to a single term. As a result, in the case of the permutation group the Wigner–Eckart theorem assumes the following form:

$$\begin{aligned}\langle[\lambda_2]r_2|T^{[N]}|[\lambda_1]r_1\rangle &= \langle[\lambda_2]r_2|[N],[\lambda_1]r_1\rangle\langle[\lambda_2]||T^{[N]}||[\lambda_1]\rangle \\ &= \delta_{\lambda_1\lambda_2}\delta_{r_1r_2}\langle[\lambda_1]||T^{[N]}||[\lambda_1]\rangle,\end{aligned} \tag{D.16}$$

where the double bars in the matrix element denote its independence of the Young tableaux r which enumerate the basis functions. Eq. (D.16) is the basis of the well-known quantum mechanical selection rule, according to which a perturbation described by a symmetric operator induces transitions in the system only between states with the same permutation symmetry.

References

[1] U. Fano and G. Racah, *Irreducible Tensorial Sets*, Academic Press, New York, 1959.
[2] E. Wigner, *Z. Phys.* **43**, 624 (1927).
[3] C. Eckart, *Rev. Mod. Phys.* **2**, 305 (1930).
[4] G.F. Koster, *Phys. Rev.* **109**, 227 (1958).

Appendix E

Second Quantization

Let us consider an arbitrary system of identical particles. In many applications the so-called one-particle approximation is used. In this approximation every particle is characterized by its one-particle wave function $\psi(\xi)$, where ξ denotes three spatial coordinates and spin projection; in Dirac notation it can be represented as a vector $|\psi(\xi)\rangle$ in the Gilbert space. It is worthwhile to note that the one-particle approximation is used in the Hartree–Fock method, on which almost all modern electron-correlation methods are based, see Ref. [1], appendix 3.

If particles are not interacting, the many-particle wave function can be represented as a product of one-particle vectors

$$\Psi_0(\xi_1,\xi_2,\ldots,\xi_N) = |\psi_1(\xi_1)\rangle|\psi_2(\xi_2)\rangle\ldots|\psi_N(\xi_N)\rangle. \tag{E.1}$$

Some of vectors in (E.1) may coincide. The many-particle wave function (E.1) can be characterized by numbers of particles in each state, or by so-called *occupation numbers*: $n_1, n_2, \ldots, n_N$. Instead of the coordinate variables $\xi_1, \xi_2, \ldots, \xi_N$ in the wave function (E.1), it can be represented as a vector in the occupation number space,

$$|n_1,n_2,\ldots,n_N\rangle. \tag{E.2}$$

The operator formalism in the occupation number space is called *second quantization*. It was introduced first by Dirac [2] for bosons; its equivalence with the Schrödinger formalism was established by Jordan and Klein [3]. The commutation

The Pauli Exclusion Principle: Origin, Verifications, and Applications, First Edition. Ilya G. Kaplan.

relations for the boson and fermion operators in the accepted modern form were formulated by Jordan and Wigner [4]. It should be mentioned that the second quantization formalism is especially important in the relativistic theory where number of particles is not conserved, see Refs. [5–7].

Let us consider the case of symmetric wave functions (the Bose–Einstein statistics) and introduce the main operators of the second quantization method.

The *annihilation operator* a_i decreases the occupation number in the i-th state by 1. By definition (below in the state vector we will indicate, except the first state, only those states on which the operators act),

$$a_i|n_1,\ldots,n_i,\ldots\rangle = \sqrt{n_i}|n_1,\ldots,n_i-1,\ldots\rangle. \tag{E.3}$$

It is evident that the only nonzero matrix element of a_i is

$$\langle n_1,\ldots,n_i-1,\ldots|a_i|n_1,\ldots,n_i,\ldots\rangle = \sqrt{n_i}. \tag{E.4}$$

The operators a_i and a_j, acting on different states, commutate

$$\left[a_i,a_j\right]_- = 0, \quad i \neq j. \tag{E.5}$$

Next introduce the operator a_i^+ which is adjoint to operator a_i. It is represented by a matrix adjoint to (E.4). The only nonzero element of this matrix is

$$\langle n_1,\ldots,n_i,\ldots|a_i^+|n_1,\ldots,n_i-1,\ldots\rangle = \sqrt{n_i}. \tag{E.6}$$

From (E.6) it follows that

$$a_i^+|n_1,\ldots,n_i,\ldots\rangle = \sqrt{n_i+1}|n_1,\ldots,n_i+1,\ldots\rangle. \tag{E.7}$$

The operator a_i^+ increases the number of particles in the i-th state by 1 and is called the *creation operator*. Similar to (E.5),

$$\left[a_i^+,a_j^+\right]_- = 0, \quad i \neq j. \tag{E.8}$$

The direct calculation, using (E.3) and (E.7), gives

$$a_i a_i^+|n_1,\ldots,n_i,\ldots\rangle = (n_i+1)|n_1,\ldots,n_i,\ldots\rangle, \tag{E.9}$$

$$a_i^+ a_i|n_1,\ldots,n_i,\ldots\rangle = n_i|n_1,\ldots,n_i,\ldots\rangle. \tag{E.10}$$

From the difference of these equations follows

$$a_i a_i^+ - a_i^+ a_i \equiv \left[a_i,a_i^+\right]_- = 1. \tag{E.11}$$

The operators, acting on different states, commute

$$\left[a_i,a_j^+\right]_-=0,\quad i\neq j. \tag{E.12}$$

Thus, the commutation relations for bosonic operators can be written as

$$\left[a_i,a_j^+\right]_-=\delta_{ij}. \tag{E.13}$$

From (E.10) it follows that the product $a_i^+ a_i$ can be considered as an operator of the particle number in the i-th state.

$$\hat{N}_i=a_i^+ a_i. \tag{E.14}$$

It is easy to prove that the commutators of $\hat{N}_i$ with a_i^+ and a_i are equal to

$$\left[\hat{N}_i,a_i^+\right]_-=a_i^+, \tag{E.15}$$

$$\left[\hat{N}_i,a_i\right]_-=-a_i. \tag{E.16}$$

Let us prove (E.15).

$$\hat{N}_i a_i^+|n_1,\ldots,n_i,\ldots\rangle=\hat{N}_i\sqrt{n_i+1}|n_1,\ldots,n_i+1,\ldots\rangle=(n_i+1)^{3/2}|n_1,\ldots,n_i+1,\ldots\rangle,$$

$$a_i^+\hat{N}_i|n_1,\ldots,n_i,\ldots\rangle=n_i a_i^+|n_1,\ldots,n_i,\ldots\rangle=n_i\sqrt{n_i+1}|n_1,\ldots,n_i+1,\ldots\rangle.$$

The difference of these equations leads to Eq. (E.15). Namely,

$$\left(\hat{N}_i a_i^+-a_i^+\hat{N}_i\right)|n_1,\ldots,n_i,\ldots\rangle=\sqrt{n_i+1}|n_1,\ldots,n_i+1,\ldots\rangle\equiv a_i^+|n_1,\ldots,n_i,\ldots\rangle.$$

The proof of (E.16) is similar.

The formalism developed above can be formulated in another form introducing so-called *ψ-operators*

$$\hat{\psi}(\xi)=\sum_i|\psi_i(\xi_i)\rangle a_i,\quad \hat{\psi}^+(\xi)=\sum_i\langle\psi_i(\xi_i)|a_i^+. \tag{E.17}$$

For some applications this presentation can be more convenient, see Refs. [8, 9].

In general case, the Hamiltonian of many-particle system can be represented via one-particle $\hat{f}_a$, and two-particle $\hat{g}_{ab}$, operators

$$\hat{H} = \sum_{a} \hat{f}_a + \sum_{a<b} \hat{g}_{ab}. \tag{E.18}$$

In (E.18) the summation is performed over all particles. The operator $\hat{f}_a$ acts only on coordinates of particle a, the operator $\hat{g}_{ab}$ belong to two particles, a and b, and describes their interaction. It can be checked, see Refs. [8, 9], that the Hamiltonian (E.18) can be represented in terms of operators a_i^+ and a_i in the following form:

$$\hat{H} = \sum_{i,j} f_{ij} a_i^+ a_j + \sum_{i,j,k,l} g_{ij,kl} a_i^+ a_j^+ a_l a_k, \tag{E.19}$$

Here, the matrix elements

$$f_{ij} = \langle \psi_i(\xi) | \hat{f} | \psi_j(\xi) \rangle, \tag{E.20}$$

$$g_{ij,kl} = \langle \psi_i(\xi_1) \psi_j(\xi_2) | \hat{g}_{12} | \psi_k(\xi_1) \psi_l(\xi_2) \rangle \tag{E.21}$$

are defined on the one-particle wave functions.

For a system on noninteracting particles, the Hamiltonian (5.19) contains only the first term,

$$\hat{H}_0 = \sum_{i_N} f_{ij} a_i^+ a_j. \tag{E.22}$$

If one-particle wave functions are taken to be the eigenfunctions of the Hamiltonian of the noninteracting system, the matrix (E.20) is diagonal and its elements are the eigenvalues ε_i of the particle energy. Eq. (E.22) can be written as

$$\hat{H}_0 = \sum_{i} \varepsilon_i a_i^+ a_i = \sum_{i} \varepsilon_i \hat{N}_i. \tag{E.23}$$

In the case of antisymmetric wave functions (the Fermi–Dirac statistics), the expression for the Hamiltonian preserves its form (E.19). However, the commutation relations for operators a_i^+ and a_i are changed [8, 9]. The commutators, which are valid for the boson operators, are changed into the anticommutators for the fermion operators. Namely,

$$\left[a_i, a_j^+\right]_+ = a_i a_j^+ + a_j^+ a_i = \delta_{ij}, \tag{E.24}$$

$$\left[a_i, a_j\right]_+ = \left[a_i^+ a_j^+\right]_+ = 0. \tag{E.25}$$

Eq. (E.25) is also valid for $i = j$, that is, the anticommutator $[a_i^+ a_i^+]_+ = 0$ or

$$(a_i^+)^2 = 0. \tag{E.26}$$

One state can be occupied by only one fermion, in full accordance with the original formulation of the Pauli exclusion principle.

The expression for the operator of the number of the fermion particles, $\hat{N}$, is the same as for bosons,

$$\hat{N}_i = a_i^+ a_i. \tag{E.14a}$$

References

[1] I.G. Kaplan, *Intermolecular Interactions: Physical Picture, Computational Methods and Model Potentials*, John Wiley & Sons, Ltd, Chichester, 2006.
[2] P.A.M. Dirac, *Proc. Roy. Soc. Lond. A* **114**, 243 (1927).
[3] P. Jordan and O. Klein, *Z. Phys.* **45**, 751 (1927).
[4] P. Jordan and E. Wigner, *Z. Phys.* **47**, 631 (1928).
[5] W. Heisenberg and W. Pauli, *Z. Phys.* **59**, 168 (1930).
[6] L.D. Landau and R. Peierls, *Z. Phys.* **62**, 188 (1930).
[7] V. Fock, *Z. Phys.* **75**, 622 (1932).
[8] S.S. Schweber, *An Introduction to Relativistic Quantum Field Theory*, Row Peterson, New York, 1961.
[9] L.D. Landau and E.M. Lifshitz, *Quantum Mechanics (Nonrelativistic Theory)*, 3rd edn., Pergamon Press, Oxford, 1977.

Index

Page numbers with n indicates footnote.

The Pauli Exclusion Principle: Origin, Verifications, and Applications, First Edition. Ilya G. Kaplan.